TASMANIA'S WILDERNESS BATTLES

TASMANIA'S WILDERNESS BATTLES

A HISTORY

GREG BUCKMAN

First published in 2008

Maps on pages xi and 214 by Ian Faulkner

Jacana Books, an imprint of
Allen & Unwin
83 Alexander Street
Crows Nest NSW 2065
Australia
Phone: (61 2) 8425 0100
Fax: (61 2) 9906 2218
Email: info@allenandunwin.com
Web: www.allenandunwin.com

National Library of Australia
Cataloguing-in-Publication entry:

Buckman, Greg, 1960–
Tasmania's wilderness battles : a history / Greg Buckman.
Crows Nest, N.S.W. : Allen & Unwin, 2008.

978 1 74175 464 3 (pbk.)

Includes index.
Bibliography.

Wilderness areas—Tasmania
Conservation of natural resources—Tasmania.
Environmental policy—Australia—Citizen participation
Economic development—Environmental aspects—Australia.
Environmentalism—Australia.

333.7820996

Typeset in 12/14 pt Centaur MT by Midland Typesetters, Australia
Printed by Ligare Book Printer

10 9 8 7 6 5 4 3 2 1

The paper this book is printed on is certified by the © 1996 Forest Stewardship Council A.C. (FSC). The printer holds FSC chain of custody SCS-COC-004233. The FSC promotes environmentally responsible, socially beneficial and economically viable management of the world's forests.

To Dick Jones, Olegas Truchanas, Peter Storey,
Peter Dombrovskis, Peter Murrell, Brenda Hean
and Max Price

Foreword

by Senator Christine Milne

> The surface of the earth is soft and impressible by the feet of men; and so with the path the mind travels. How worn and dust, then, must be the highways of the world, how deep the ruts of tradition and conformity! I did not wish to take a cabin passage, but rather to go before the mast and on the deck of the world, for there I could best see the moonlight amid the mountains. I do not wish to go below now.
>
> Henry David Thoreau, *Walden*

Tasmania is my home. It is where I ran free as a child on the farmlands and beaches of the north-west coast and where as a young adult at Cradle Mountain, I first encountered awe inspiring wilderness. Tasmania has nurtured and fed me with its bountiful beauty, clean air, clean water and uncontaminated soil and in return with thousands of my fellow Tasmanians and others from around the world, I have refused to take a cabin passage in her defence.

That Tasmania existed in an almost pristine state in the 1950s and 1960s was not because it was valued for these things, rather wilderness beyond the settlements was still seen as alien. ET Emmett, the first director of the Tasmanian Tourist Bureau, wrote in 1953, '40 miles due west of Hobart is a veritable "no man's land", where any lone traveller would be taking his life in his hands. It is an inferno of mountains, gorges and impenetrable forests'.

It is over this 'inferno' that the wilderness battles of the past half a century have been waged around kitchen tables, over back fences, in the forests, on the streets, in the Parliament, in board rooms, court rooms and back rooms. The battle lines of the mind were drawn between those who valued and identified with wild, untamed places and those who saw only a resource to be dug up, cut down, drowned, shot, roaded, dumped into and built upon.

Because the wilderness battles have been relentless and ongoing, from Pedder to the Franklin, from the Lemonthyme to Farmhouse Creek, the Styx Valley and the Tarkine, from North Broken Hill at Wesley Vale to Gunns in the Tamar Valley, there has been no time to record the history, to celebrate the victories, to grieve for the losses and to pay tribute to those who have gone before the mast and who continue to see the moonlight amid the mountains for us all.

Greg Buckman's book is a long overdue story of Tasmania's wilderness battles. It is the history that needed to be written so people can know and be inspired by the Tasmanian wilderness and its champions. What he describes as an environment movement is a group of some of the most selfless, brave and committed people I have ever known. I am grateful to have spent the past thirty years of my life on the journey of environmental activism with them. Their stories still need to be written and I hope that this book inspires many others as the experiences of campaigners long hidden in dusty garages and in the recesses of the mind bubble up to inform future generations of the legacy of those on whose shoulders they stand.

This book of battles is not without end. We are a small island and our wilderness forests and protected areas remain under siege. There is such a thing as being too late. I hope that this book throws an arm over the shoulder of the movement, warms its heart and strengthens its resolve to stand tall on the deck of the world. I hope it inspires you to join us to save Tasmania's remaining wilderness.

Contents

Figures

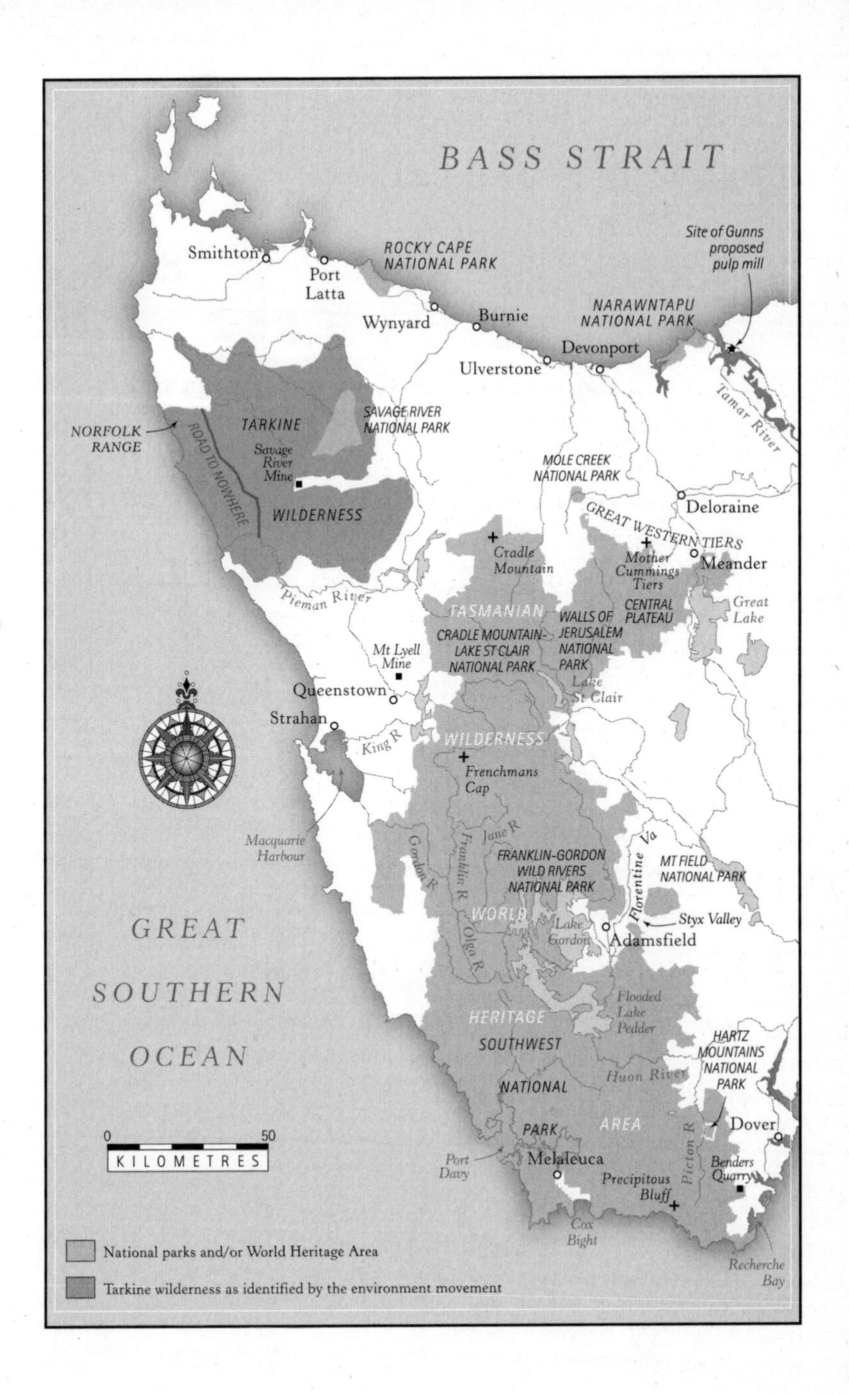

BASS STRAIT
Smithton
Port Latta
ROCKY CAPE NATIONAL PARK
Site of Gunns proposed pulp mill
Wynyard
Burnie
NARAWNTAPU NATIONAL PARK
Devonport
Ulverstone
Tamar River
NORFOLK RANGE
ROAD TO NOWHERE
TARKINE
SAVAGE RIVER NATIONAL PARK
Savage River Mine
WILDERNESS
MOLE CREEK NATIONAL PARK
Deloraine
GREAT WESTERN TIERS
Cradle Mountain
Mother Cummings Tiers
Meander
CENTRAL PLATEAU
Great Lake
Pieman River
TASMANIAN
WALLS OF JERUSALEM NATIONAL PARK
CRADLE MOUNTAIN-LAKE ST CLAIR NATIONAL PARK
Mt Lyell Mine
Queenstown
Lake St Clair
Strahan
King R
WILDERNESS
Frenchmans Cap
Macquarie Harbour
Gordon R
Franklin R
Jane R
Florentine Va
FRANKLIN-GORDON WILD RIVERS NATIONAL PARK
MT FIELD NATIONAL PARK
WORLD
Lake Gordon
Styx Valley
Adamsfield
Olga R
GREAT SOUTHERN OCEAN
Flooded Lake Pedder
HERITAGE
SOUTHWEST
HARTZ MOUNTAINS NATIONAL PARK
Huon River
NATIONAL
PARK
AREA
Dover
Picton R
0
50
KILOMETRES
Port Davy
Melaleuca
Benders Quarry
Precipitous Bluff
Cox Bight
Recherche Bay
National parks and/or World Heritage Area
Tarkine wilderness as identified by the environment movement

Preface

It is often said that to know where you are going you need to know where you have come from. This is particularly true of Tasmania's wilderness battles. But despite the drama and poignancy of the battles over the state's wild areas there is little documented history of them. This has led to a situation where, in the words of Lake Pedder activist Kevin Kiernan, few people involved in the battles 'seem to recognise any history before their own'.[1] This book fills that void. It is an accessible summary of all Tasmania's wilderness confrontations. It describes what battles have taken place and why they have taken place.

The events in the book are largely seen through the eyes of those in the environment movement. This does not mean, however, that I avoid criticism of the movement. There have been several occasions when the movement has mis-stepped (as well as many when it has trodden skilfully) and I have tried to be honest about its shortcomings.

The book is arranged thematically with different chapters for hydro, forestry and mining development as well as two chapters that outline the history of national parks in Tasmania. This structure accentuates the threads that run through different types of wilderness battles. The book covers wilderness battles throughout all of Tasmania but focuses on those of the south-western part of the state: the greatest temperate wilderness in Australia.

The account stands on the shoulders of many activists and writers who have gone before it. It would not have been possible unless concerned people in Tasmania had been prepared to stand up against the state's powerful resource development interests. A huge thanks is due to all those brave souls, especially those in The Greens, The Wilderness Society and the Parks and Wildlife Service with whom I have spent many great times. I would like to extend particular thanks to Helen Gee, Chris Harries, Greg Middleton, Geoff Law, Kevin Kiernan, Bob Brown, Simon Alston and Adam Burling for their generous sharing of their inside knowledge and checking of my text and to Christine Milne for writing the foreword. A major thanks is also due to Louise Egerton from Allen & Unwin for showing faith in the publication and being a most supportive editor, to Rebecca Kaiser for nursing it through production and to Karen Ward for her superb copy-editing. One of the delights of writing the book was bringing to light, again, the extraordinary knowledge contained in several university theses and occasional papers that deal with Tasmania's wilderness. My great appreciation goes to Debbie Quarmby, Louise Mendel, Gerard Castles, Phillip Hoysted, Ron Sutton and Jayne Balmer for making this possible. The book has been lifted by the inclusion of Ron Tandberg's great Tasmanian wilderness cartoons; many thanks to him for giving me kind permission to use them. I am also indebted to those who helped with image selection and processing, especially Dianne Mapley, Jill McCulloch, Margaret Steadman, Fiona Preston, Caitlin Carew, Jim England, Liz Dombrovskis, Paul Oosting, Peter Cameron, Richard Bell, Patsy Jones, Margaret Ambrose, Suzi Pipes, Glenys Tandberg, Tim O'Loughlin, Wilf Elvey, Graeme Harrington, Chris Roach and the diligent staff of the Archives Office of Tasmania.

My biggest thank you, by far, goes to my wonderful partner, Katrina Willis, who provided invaluable focus for the book and to my parents, Jill and Ted Buckman, who provided more support than they will ever realise.

INTRODUCTION

The meaning of wilderness

Wilderness is as much a state of mind as it is a physical reality. Humankind needs wilderness for peace of mind, for release and to make sense of the largely artificial world it has created for itself. US philosopher Richard Jeffrey once said: 'The hours when the mind is absorbed by beauty are the only hours we truly know'. When the mind is steeped in the spirituality of wilderness it is capable of assuming the type of completeness and connectedness suggested by Jeffrey. Wilderness is a church where you only need believe in the beauty of the natural world around you. As the number of people in the world grows, and as development spreads, we need wilderness more than ever. Another US philosopher, Henry David Thoreau, captured the importance of wilderness to us all when he said: 'in Wildness is the preservation of the world . . . in short, all good things are wild and free'. As humans build ever-larger castles of commerce, government and consumption we need ever-stronger bridges to the serenity of wilderness that lies beyond the castle moats.

Yet despite our increasing need for wilderness, it is in retreat. In 1700, before the Industrial Revolution got under way, there were about 3.5 billion hectares of forest in the world but today there is only about 2 billion hectares left.[1] We are destroying our church. If we do not stop soon, wilderness will become a museum piece: a long-forgotten concept that future generations will only know about by reading history books. Founder of the Friends of the Earth conservation group, David Brower, said of the fragility of wilderness:

> There aren't many places left where we, ourselves, can choose to exploit or leave wild. Although the budget of natural things may have looked unlimited to grandfather, we know it is a finite budget. Wilderness is a fragile thing. People can break it but not make it. And we are quite capable, in our own time, of breaking it all . . .[2]

Despite Brower's warning, humans are breaking wilderness. Fortunately, some have realised that we have to act if we are to retain what remains of our wild lands. Since the middle of the twentieth century there has been a growing resistance to the plunder of natural areas. But as the resistance gains strength so, too, do the forces pitted against it. We are becoming more aware of wilderness but are firing increasingly sharper arrows against it. This contradictory attitude is more evident in Tasmania than in any other part of Australia.

In Tasmania wilderness imagery abounds: it is in tourist ads, it sells cigarettes, it sells beer and the state government's logo features a thylacine peering through some wild growth. But the state also has ever-present reminders of the threats to wilderness: log trucks constantly ply the state's highways, there are large dams in every major region of the state linked by long lines of transmission towers, and a drive through the state's west is a constant reminder of how pervasive the effects of the state's mining industry have become. In most parts of Australia the clash between humans and wilderness is at some remove from daily life but in Tasmania it is in your face. Given this, it is far from surprising that the state has a long history of hosting some of Australia's most passionate wild area battles.

For over 60 years Tasmanian environmentalists have fought the might of the state's forestry, hydro and mining interests. Before the Second World War, resource developers faced no environmental resistance to their despoliation of wilderness areas but they often left large parts of the state untouched, particularly in the west.

Since the war they have faced much more resistance but have also tried to reach further into the state's unspoilt areas than ever before.

The following chapters chronicle the emergence of resource development in Tasmania and its confrontations, over time, with the state's wilderness defenders. The two have often had brutal and bloody clashes and have always had vastly different views of the meaning of wilderness. To developers it may be intriguing but has nearly always been expendable; to environmentalists it is sacrosanct and its destruction epitomises the shallowness of much of human progress. Wilderness is variously seen as a source of spirituality or a business opportunity, but in essence it remains elusive to humankind, particularly in Tasmania. Perhaps, ultimately, there is no one meaning of wilderness. Perhaps it is meant to be different things to different people. Tasmanian photographer Peter Dombrovskis once said: 'nature in a sense is already sufficiently unknowable that it is sufficient to present it as it is'.[3] Maybe one can never fully understand wilderness and our struggle with it reflects our internal struggles as much as anything else. The following chapters describe how this struggle has played out in Tasmania.

HYDRO

I

Growth of the HEC

By the 1950s one word summed up Tasmania: electricity. By that decade Tasmania had the second highest per capita consumption of electricity anywhere in the world,[1] its power demand was doubling every five years and a new hydro-electricity generation station was opening nearly every year. Hydro power ruled Tasmania; there seemed no limit to the engineering feats it could perform, nor the power-hungry industries it could attract. By the 1950s hydro electric development in Tasmania had come a long way in a short time: just 40 years earlier there had been no centralised hydro development in the state. But just 40 years later, in the 1990s, the last concrete would be poured into the last hydro dam that will probably ever be built in Tasmania.

What sets hydro development apart from other types of resource development in Tasmania is its technological wizardry. No matter how cynical one might be, it is hard not to be moved by the power of hydro schemes to divert whole rivers through mountain ranges and their ability to hold back enormous volumes of water. Faith in hydro power is faith in technology: a faith that says that technology can, and will, find answers to humankind's challenges. In Tasmania, hydro electric development embodied society's faith—and subsequently its pessimism—about technology like no other industry.

Early hydro development

None of the mid twentieth century confidence in the benefits of hydro electric development in Tasmania was evident at the start of the century. Some use was being made of water power but not a lot. Most flour mills throughout the state were using hydro power but few other industries were. Launceston City Council changed all that, turning Tasmania into a hydro trailblazer when it established the state's first hydro electric power station at Duck Reach, on the South Esk River, in 1895. This was just two years after the world's first hydro station was opened in Canada at Niagara Falls. Given the numerous rivers that run throughout the state, it was inevitable that Tasmanian hydro development would not end with Duck Reach.

The idea of a more comprehensive harnessing of the state's water power was taken up by Scotsman James Gillies. In 1908 he suggested to the state government that the hydro potential of the Great Lake, located in the state's Central Plateau, could be utilised. He argued the lake could supply electricity to an energy-intensive metallurgical plant he was interested in establishing. Since the scheme would generate electricity equal to three times the electricity demands of Hobart, a major energy-hungry industry such as Gillies' plant was a crucial part of the scheme's justification. In later years, the Hydro Electric Commission (HEC), as well as a succession of Tasmanian governments, was criticised for being obsessed with generating electricity for power-hungry industries, but it is important to recognise that this die was cast at the start. Hydro electric development has always been promoted in Tasmania as a means of attracting new industry and was never justified as a way of meeting existing electricity demand.

The Tasmanian government liked Gillies' idea but insisted the Great Lake scheme be privately developed and gave him four years from 1909 to build it.[2] Like later hydro developments in Tasmania, the scheme involved the flooding of a pristine environment: the Great Lake was a large body of water that dominated the state's Central Plateau. The area was not as wild and untouched as the state's south-west wilderness but it still had a significant natural beauty. The Great Lake was a magnificent lake to behold: it was 27 kilometres long and 13 kilometres wide with many small bays and inlets along its 157-kilometre shoreline. The lake was also fairly shallow; never more than 7 metres deep, with five islands protruding above its surface.

Gillies' vision soon ran into trouble. To keep within his four-year time limit he had to build throughout the punishing Tasmanian highland winters. Also, like many subsequent hydro schemes in the state, his scheme proved much more expensive to build than he predicted. Despite Gillies being given two deadline extensions, he got into ever deeper financial trouble, compounded by a scarcity of overseas capital during the prelude to the First World War. Gillies' situation became so dire that in 1914 the state government bought him out and a new Hydro Electric Department took over his hydro scheme. As many Tasmanian businessmen after him would do, Gillies made sure his proposed energy-hungry business (which had gone from being a zinc plant to a carbide works) retained its access to a cheap and plentiful source of electricity by securing a commitment that he would be supplied with cheap hydro power despite his sale of the Great Lake scheme.[3]

The chief engineer for Gillies' scheme, John Butters, became the head of the new Hydro Electric Department. Within two weeks of his appointment Butters formalised the relationship between hydro electric development and energy-intensive industry by writing to Premier John Earle suggesting that Tasmania establish an electricity subsidy system. He said it would allow the state to attract new industries by offering them cheap hydro power that would undercut electricity prices in other states.[4] This strategy became central to hydro development in Tasmania thereafter. What the state lacked in economic attractiveness, it would make up for in bargain basement electricity prices. The Tasmanian government was enthusiastic about Butters' power subsidy plan and encouraged him to develop it. In 1915 the Hydro Electric Department signed its first bulk power contract with Gillies' carbide company, which would operate at Electrona, south of Hobart. At roughly the same time, the federal government began offering financial inducements for the establishment of a zinc factory in Australia following a shortage of the metal during the First World War. This eventually led to the signing of a hydro bulk power contract with the Amalgamated Zinc company in 1916, the same year that Waddamana, the power station that serviced the Great Lake scheme, was finally opened.[5]

The wisdom of attracting energy-hungry industries to Tasmania with inexpensive power was always contested; there was unease about it even in 1916. There was significant press criticism about the terms agreed on to secure Amalgamated Zinc's smelter. To maximise the chances of the smelter going ahead, the Tasmanian government offered the company a low-cost lease of land in Hobart and the Hydro Electric Department offered electricity at rock-bottom prices in what became known as 'the zinc bargain'.[6] It was even suggested that the power was being offered at such cheap prices that the smelter might end up swallowing all of the state's power output (which ended up not far wrong: by the early 1930s it was consuming three-quarters of the Hydro Electric Department's electricity output). The Tasmanian government defended the arrangement by arguing it would never happen again and that in future it—not the Hydro Electric Department—would negotiate the power deals (which never happened). Undeterred by the controversy, the proponents went ahead and plans were finalised in 1919 to expand the Great Lake scheme after more energy-intensive businesses expressed interest in setting up in the state. In what was to set a major precedent, when the tender was advertised for the expansion of the scheme, all of the private sector bids were deemed too expensive so the Hydro Electric Department undertook the work itself,[7] confirming its status as both a power-generating and a dam-construction agency.

The Great Lake scheme enlargement inundated most of a unique feature known as the Shannon Rise. This was half a kilometre of constantly flowing cold water; a consequence of the original Great Lake scheme. It was an ideal hatching ground for a moth that drew a lot of trout to the area and, therefore, eager fishermen. The

Shannon Rise was far from a wilderness area but was a significant site that the Hydro Electric Department showed scant interest in preserving.

Most of the major mistakes of Tasmanian hydro electric development began in its early years. The zinc bargain was such a mistake. Yet another early mistake was the 1929 decision to make the Hydro Electric Department a commission: a semi autonomous body with significant independence from government.[8] During the Lake Pedder and Franklin River controversies the HEC was often portrayed as a law unto itself and in 1978 the HEC even successfully agitated for the removal of a minister it did not like.[9] Once the commission structure was in place, the government had no ability to direct it apart from approving its borrowings. In the 1920s and 1930s the HEC's authority was further enhanced by its takeover of the state's electricity distribution systems (which had hitherto been controlled by councils). Successive state governments had blind faith in the independence of the HEC and the 1929 decision to make it a commission was one that some later governments would regret.

Expansion in the 1930s

The Great Depression hit Tasmania hard: in 1928 unemployment was about 11 per cent and by 1931 had leapt to 27 per cent.[10] Amidst this hardship, state elections were held in 1934. Voters were desperate for job-creation schemes to combat the state's economic woes. The man who seemed to have the answers was the leader of the Labor opposition, Albert Ogilvie. He saw resource development as the way out of the state's economic problems. He proposed the construction of a new hydro scheme, to be built in an area called Tarraleah. Ogilvie's scheme would utilise water from the upper Derwent River and Lake St Clair and would be significantly larger than the Great Lake scheme, creating more than

1000 construction jobs. Unlike the Great Lake scheme, however, the Tarraleah proposal would have a direct impact on a wilderness area by utilising the waters of Lake St Clair in the Cradle Mountain–Lake St Clair National Park, necessitating the construction of a small weir at one end of the lake. The resultant rise in water level would flood the lake's shoreline beaches. Although not on the same scale as the later flooding of Lake Pedder, this construction nonetheless marked the start of hydro development impacting on the state's wilderness areas. The flooding of Lake St Clair was opposed by the national park management authority of the time, the Scenery Preservation Board, but it was powerless against the all-conquering HEC (see Chapter 9).

Many people see the 1960s and 1970s Labor premier of Tasmania, Eric Reece, as the 'Father of the Hydro' ('the Hydro' being the HEC's colloquial name), but the title really belongs to Ogilvie. For him, job creation through hydro schemes was entirely consistent with Labor principles: he and many of his cohorts even saw it as a form of electric socialism. Ogilvie was probably influenced by a new fad for building large dams that was sweeping the world at the time. The mania was spurred on by the construction of the 220-metre-high Hoover Dam across the Colorado River in the United States. Completed in 1935, the dam became a renowned symbol of humankind's ability to tame nature and build its way out of the Depression.

Voters liked what Ogilvie said and his Labor government was elected in 1934. His much-touted Tarraleah scheme began operating in 1938. As if to confirm Ogilvie's optimism about the potential for cheap hydro power to generate jobs, between 1935 and 1938 several publishers expressed interest in building a newsprint mill in the Derwent Valley, near Hobart, through a new company: Australian Newsprint Mills (ANM). In 1936, Associated Pulp and Paper Mills (APPM) also signalled interest in establishing a pulp and paper mill at Burnie, in the north of the state (the only other paper mill in Australia at the time was one in

Victoria). In a portent of later standard practice, when potential competitors to APPM questioned the favourable power prices it was offered, the HEC flexed its muscles of independence and refused to reveal them.[11]

Ogilvie died of a heart attack in 1939 and the premiership passed to Robert Cosgrove, who continued his support for hydro development. Cosgrove, however, was less in awe of the HEC's management style than Ogilvie was and in 1941 unsuccessfully sought to reinstate it as a government department.[12] In the late 1930s and early 1940s it was also becoming apparent that the HEC's electricity supply was not keeping up with ever-growing power demand so the generating capacity of the Great Lake and Tarraleah schemes was expanded.

Just as the need for zinc in the First World War had significantly increased the state's hydro development, the shortage of another strategic metal, aluminium, further increased the demand for hydro power in Tasmania after the Second World War. During the early part of the war, the federal government expressed interest in establishing an aluminium plant in Australia. This became concerted action after Japan entered the war. It takes a lot of electricity to produce aluminium—it is sometimes known as 'congealed electricity'—so Tasmania's cheap bulk power prices always made it a frontrunner for the smelter.

In September 1943 Prime Minister John Curtin announced that an aluminium smelter would indeed be built in the state. Tasmania was not alone in using hydro electric power to attract aluminium smelting; throughout the second half of the twentieth century many hydro electric powered regions of the world did the same thing. Aluminium smelters were established in Iceland, Norway, Canada and New Zealand, all of which used the inducement of inexpensive hydro power to attract them. In 1944 the Tasmanian and federal governments signed an agreement to jointly fund the smelter's £3 million construction cost. After the war Tasmania was offered a confiscated Norwegian aluminium smelter as part of German war reparations but it was old and inefficient so work began on a new smelter at Bell Bay, near Launceston.[13] While the smelter had no immediate impact on the state's wilderness areas, it went on to have a huge indirect impact by significantly increasing the state's power demand and therefore the need to develop more rivers.

Although the HEC was able to cope with a doubling of state power demand between 1930 and 1940, it was obvious by the end of the Second World War that it would not be able to cope with an impending increase in industrial power demand without another huge expansion. Cosgrove called for a doubling of the HEC's capacity and the HEC made the then daring suggestion that its workforce be expanded by recruiting thousands of migrant workers to work on all its new schemes. The state government took up this suggestion. As long as the HEC could keep building new dams its postwar future looked bright. The HEC felt equal to the task and was so infused with optimism that the head of the commission, Alan Knight, told a 1947 parliamentary tour that in twenty years' time it would have five times its generating capacity.[14]

The HEC's confidence was buttressed by a new 1944 *Hydro Electric Commission Act* that increased its independence. The one check on the powers of the HEC that both its 1929 and 1944 acts included was the need for parliament to approve its raising of loan funds. But even this check was ineffective as a review mechanism because the HEC never gave parliament alternatives to the schemes it was proposing (until its Franklin scheme). It always asked parliament to approve its new borrowings on a take-it-or-leave-it basis. One-time deputy premier, Kevin Lyons, observed:

> . . . when a new scheme comes forward parliament has the right to accept it or reject it. But when any new scheme has been proposed, no alternative scheme has been presented to parliament to consider. The question then is of either accepting or rejecting the scheme suggested.[15]

The postwar push

The new aluminium smelter went on to become Tasmania's largest electricity consumer. Inevitably, the price it paid for its power attracted attention, just as the zinc smelter contract did. The terms of the Tasmanian government's agreement with the federal government obliged the HEC to offer power to the aluminium smelter at a rate that was interpreted to mean at cost.[16] The HEC and the aluminium company could not agree on a price and negotiations between them became heated and drawn out. Finally, a 25-year agreement was signed in late 1948 in which the smelter succeeded in getting very low prices.[17] Tellingly, the following year the Commonwealth Grants Commission suggested that the HEC should raise its prices but the commission refused to do so, saying it would scare off new customers.

To meet the heightened postwar power demand the HEC began work on two new schemes: one at Trevallyn, on the Tamar River near Launceston, and another called Tungatinah located next to the Tarraleah scheme. To keep ahead of the rapidly rising demand for power, by the end of the 1940s no less than six new schemes were under construction. In 1951 three new power schemes on the Derwent River (which flows into Hobart) were approved. By 1952 the HEC was spending more on its works program in one year than it had spent in all of the preceding 35 years. Tasmanian hydro development was unstoppable.

When the HEC started developing the Derwent River it signalled an interest in developing rivers in both the eastern and western parts of the state that would eventually bring it into sharp conflict with the environment movement. In the early 1950s the head of the HEC, Alan Knight, ordered the first detailed investigations of the Gordon, Franklin, Arthur, King, Huon and Pieman rivers, all of which flowed through the west of the state. In 1953 the HEC even placed flow recorders on the upper reaches of the Gordon River, well before there was any public inkling of hydro plans for the state's south-west wilderness.[18] At the time most of south-west Tasmania was remote and little was known about it. Investigation of its hydro potential therefore included the first detailed mapping of the area, an exercise made easier by the introduction of helicopters for HEC survey work.

Although the HEC's plans to develop the Gordon River, which included the flooding of Lake Pedder, went on to become the first major clash between proponents of hydro development and the conservation movement, a foretaste of the battle came when the Trevallyn power scheme began operating in 1955. The scheme decreased the flow of the South Esk River through Launceston's scenic Cataract Gorge by as much as two-thirds. Once the impact was known there was a backlash. The Launceston *Examiner* ran a headline: 'The Beauty Spot that Disappeared'. A Launceston member of the Legislative Council, the state's upper house, Jack Orchard, said: 'Launceston's most beautiful tourist asset was prostituted to the production of power'.[19] As a face-saver the HEC shared the cost of building a series of weirs in the gorge that created the impression of greater flows but the issue remained a sore point.

By the mid 1950s the HEC's financial situation began catching up with the juggernaut of its new developments, forcing a series of controversial domestic and commercial electricity price increases. Despite the price hikes, state government spending on health, education and infrastructure had to be reduced to meet the ongoing financial hunger of the HEC. Along with some power rationing, the price increases generated some scepticism about the future of hydro industrialisation, but not enough to slow the HEC's pace. In 1957 the commission unveiled its most technologically ambitious scheme so far: a doubling of the generating capacity of the Great Lake through the construction of its first underground power station. The following year the HEC gave notice of equally bold power developments on the Mersey and Forth rivers and—ominously—the Gordon, Franklin, Pieman and King rivers in western Tasmania.

In 1958 the premier most associated with hydro development in Tasmania, Eric Reece, took over from the ailing Cosgrove. Confirming its undiminished march, in the summer of 1960–61 the HEC set up major investigation camps at the future site of a dam it had planned for Lake Pedder and at the junction of the Gordon and Franklin rivers where it hoped to build another dam. The need to build new hydro schemes in the wilderness of western Tasmania was closely linked to the large electricity appetite of the Bell Bay aluminium smelter. In 1958 the Tasmanian government urged the federal government to double the capacity of the smelter, to which it agreed on condition that the capital was sourced from the private sector.[20] The federal government then sold its two-thirds share to the Comalco company (although the Tasmanian government retained its share). In 1962 Comalco reached agreement with the HEC on a power price that would enable the company to treble its output by 1964.[21] At the same time, work on the new Derwent River power schemes was nearing completion and work soon began on another three power stations on the same river. Nothing, it seemed, could stand in the HEC's way.

2

Lake Pedder

Until the early 1960s Tasmanians held the Hydro Electric Commission (HEC) in high regard. Although there had been minor controversies—including the 'zinc bargain' of 1916 and the withdrawal of water from Launceston's scenic Cataract Gorge in the 1950s—in the minds of most the HEC had delivered jobs and money to the state and could do no wrong. But this exalted status did not last. The HEC was eventually brought down to earth by the battle for a majestic lake in the heart of the state's south-west wilderness: Lake Pedder.

Lake Pedder was a unique 5-square-kilometre body of water that had a spectacular 3-kilometre quartzite sand beach that took the better part of an hour to traverse. The lake was flanked on three sides by a curtain of precipitous mountains; it was a place apart. Artist and Pedder activist Max Angus said:

> No description, however detailed, could remotely convey the sense of awe and wonder felt by those who saw this magic place . . . the overwhelming sense is of space and light . . . you look at an infinity of sky and mountains reflected all around in the impeccable surface of the lake and feel disbelief that you are standing on a beach nearly a thousand feet above sea level.[1]

Lake Pedder was the jewel of Tasmania's south-west wilderness, a jewel untouched by any development until the HEC came along. After the Second World War it became popular with bush-walkers. It became so popular that the Hobart Walking Club was instrumental in having it declared a national park in 1955, making it the second national park in the south-west after Frenchmans Cap National Park, created in 1941.

The announcement of the Lake Pedder scheme

The excision of part of Mount Field National Park to feed the forest resource hunger of Australian Newsprint Mills in the 1940s (see Chapter 4) had demonstrated that national parks were not necessarily safe. Nevertheless, Tasmanians assumed Lake Pedder was. With regard to the development possibilities of the lake, just before Lake Pedder National Park was created the HEC had even said: 'the commission does not consider this development to be likely in the next twenty years and therefore raises no objection to the proclamation of the reserve'.[2] It was with utter disbelief, then, that Tasmanians received the announcement on 10 December 1962 by their ardently pro-hydro premier, Eric Reece, that the Gordon River, which flowed through the middle of the south-west, would be investigated for its hydro development potential. The HEC was interested in building no less than three separate hydro schemes on the river, the middle one of which might affect Lake Pedder. Reece (who became popularly known as 'Electric Eric') and the HEC were vague about details and emphasised they were only interested in investigations. The day after the announcement, the HEC said it would build a 64 kilometre jeep track through to the junction of the Gordon River and the Serpentine River, that flowed out of Lake Pedder, again emphasising that it was all being done in the name of research, nothing more.

While the HEC's plans were imprecisely worded, Reece's announcement had major implications. Hydro schemes had never been built in the west of Tasmania before and everyone knew it was a new frontier for the HEC. No roads then penetrated the south-west wilderness. The nearest roads were one that ran to Maydena, near Mount Field on the eastern side of the wilderness (a day's walk from Lake Pedder) and the Lyell Highway that ran through the northern edge of the area. The HEC's jeep track would open the region up in a big way. The HEC was staking a major claim to Tasmania's most significant wilderness area—there would be no turning back. In 1964 it upgraded the jeep track to a graded road after being given £2.5 million by the Menzies federal government. The HEC kept emphasising the benign nature of its interest but its investigations looked more and more serious. Surveyors placed several flow recorders on the Serpentine and Gordon rivers and undertook detailed investigations of the area. Increasingly, the HEC's intentions seemed dubious and further questions were asked by the environment movement, but Reece and the commission kept silent. Reece had no doubt about the virtue of the schemes; for him, untamed rivers like the Gordon represented 'wealth flowing into the sea'.[3]

Although unaccustomed to environmental battles, Reece knew he would have a fight on his hands over Lake Pedder so he set about finding an olive branch for the development. Reece thought if he could announce a major reservation of a large part of the south-west at much the same time as he announced the Lake Pedder hydro plans he could take the sting out of the controversy. In August 1964, he told the media he was setting up an interdepartmental committee to investigate the reservation possibilities of the south-west. There was a lot of unprotected and unallocated land in the area and Reece saw potential for declaring a large new national park that might give him cover. Without any apparent sense of irony, Reece even said more reservation in the south-west would mitigate against 'gross and wanton destruction' and would 'protect the region against undue damage'.[4]

A conservation group that had been created in 1962 by walking clubs, field naturalist groups and the Youth Hostels Association—the South West Committee—took a major interest in the intrigue surrounding Lake Pedder and asked to be included on Reece's reservation committee. Its request was rebuffed by the premier, who restricted his committee's membership to the HEC, the Surveyor-General, the Forestry Commission and the Mines Department.[5] Such restricted membership made it clear that Reece intended to avoid reserving areas of wilderness that might one day be ripe for development. The only conservation-minded committee representative was the Scenery Preservation Board, forerunner to the state's National Parks and Wildlife Service, which by then had an unassertive attitude to preservation (see Chapter 9).

One of the great legacies of the Pedder controversy was its radicalising influence on the Tasmanian environment movement. Although the South West Committee was made up of many concerned and altruistic people, it was not a bold organisation. Instead of taking its message to the people, the South West Committee invariably opted to work through bureaucratic channels. This was the favoured modus operandi of most Australian environment groups at the time. It drew up maps of its preferred reservation of the south-west and made submissions to hearings about Lake Pedder, but was generally shy of going public. Another major weakness was its assumption that the flooding of Lake Pedder was a fait accompli and its focus on securing an enlarged national park for the area instead of trying to save the lake.[6] This was a tactical mistake that later Tasmanian environment groups would learn from.

As the HEC's investigative work progressed, and as Reece's interdepartmental committee began drawing up reservation plans that assumed Lake Pedder would be drowned, there was less and less room for evasiveness. Finally, in June 1965, Reece admitted that 'there may be some modification' to Lake Pedder.[7] The worst fears of the environment movement were realised. Many Tasmanians were horrified. Despite the demand for more details, Reece would say no more. In 1966 he secretly sought approval from the Menzies government for a $47 million grant towards the Pedder scheme, funds which he did not disclose until the scheme's legislation was passed by state parliament the following year.[8]

Another clandestine operation undertaken at the time was a scientific expedition to the lake in February 1967. Relatively little was known about the lake's environment so the HEC allowed scientists from the Hobart and Launceston museums to spend a fortnight there. But the investigations were compromised by the exclusion of any botanist, and therefore of any plant surveys, and also by the HEC's disallowing the group from independently publishing its own report.[9]

By early 1967 there was widespread angst about the future of Lake Pedder and, finally, on 25 May the HEC's official report on the scheme was tabled in parliament. Like previous HEC proposals, it contained no alternatives: just its preferred option. In the words of Pedder activist Kevin Kiernan, it read 'like a sales brochure'.[10] But the truth was out and the full extent of Reece's 'some modification' euphemism was revealed: Lake Pedder would be drowned below a new lake nearly 250 square kilometres in area. A dam would be built on the Serpentine River flowing out of Lake Pedder, as well as on the nearby Huon River which, together with a third dam, would completely flood the Lake Pedder valley. To the HEC this would enhance the original lake and it wanted to keep the name of Lake Pedder for the new impoundment, a move that conservationists saw as deliberate obfuscation by the HEC designed to make people feel they had a bigger, better Lake Pedder. To this day many conservationists refuse to use the name. To Reece the scheme represented development of worthless land that only contained 'a few badgers, kangaroos and wallabies, and some wildflowers that can be seen anywhere'.[11] As far as the HEC was concerned, its new lake would be as beautiful as the original one, possessing 'almost all of the scenic attractions of the existing Lake Pedder'.[12]

The HEC's new Lake Pedder would be one of two new lakes that together would make up its 'Middle Gordon' scheme. The other lake was to be called Lake Gordon and would be created by damming the Gordon River. It would be connected to Lake Pedder via a narrow channel. The combined storage of the new

Lake Pedder and Lake Gordon would equal eighteen times the volume of Sydney Harbour. The month after its report was tabled the HEC opened its road into the Lake Pedder area and no one was surprised to find it conformed to the road infrastructure the scheme required. Kiernan would later memorably say: 'I saw my temple ransacked'.

The HEC's report induced a huge public outcry and a petition with over 10 000 signatures was organised by the conservation movement. In March 1967 a new group was created to defend the lake, this one called the Save Lake Pedder National Park Committee. Unlike the South West Committee, this group was outspoken and immediately began to raise public awareness by organising slide shows, public meetings and publications.[13] Between 1962 and 1976 no less than five environment groups were created to defend south-west Tasmania. This turnover was part of an evolution of the Tasmanian conservation movement that allowed it to achieve new levels of assertiveness. It was a necessary part of the growth of the movement during this formative time. There was significant tension between older activists, who were used to keeping quiet in a close-knit community like Tasmania's, and younger ones who were less worried by local sensibilities. The younger activists ultimately succeeded in creating more vibrant conservation organisations but for a long time felt silenced by influential elders. It was easier to create new organisations than to try to radicalise established groups. Each new group generally had an informal structure so few resources were needed in creating the new organisations.

At the same time as the tabling of the HEC's Lake Pedder scheme report, Reece began to get the new south-west national park in place. In April the interdepartmental committee gave its report to the Minister for Lands. It recommended the creation of a large new national park that would extend from the flooded Lake Pedder through to the south coast and would take in iconic mountain ranges like the Arthur and Anne ranges. By implication, however, it backed the HEC's plans and carefully avoided recommending the reservation of any areas that might have significant mineral or forestry resources, such as the Port Davey area on the west coast or the Precipitous Bluff area on the south coast.

The Legislative Council inquiry

To outflank public unrest Reece hoped to rush the legislation authorising the Pedder scheme through parliament, but when the HEC's report was tabled the South West Committee said it was 'deeply concerned by the absence of impartial examination of the very important factors involved'. The upper house of the state parliament, the Legislative Council (mainly made up of independent members), had similar feelings and appointed its own committee of inquiry. Reece was determined not to be stymied by the inquiry and introduced his legislation before it could get under way. Kiernan said the upper house inquiry was 'kept in the dark' and did little more than 'grumble' about the Pedder scheme.[14] But the inquiry was, nonetheless, somewhat open to the conservation movement's point of view and was interested in whether any alternatives to the scheme had been investigated. The HEC managed to convince the inquiry that coal and nuclear alternatives were not viable and went on to argue that the steep rises in electricity demand that Tasmania was experiencing meant its recommended scheme was the only sensible option.[15] What the inquiry was stunned to learn, however, was that the HEC had formulated and costed alternatives to flooding the lake that involved making the dam that held back the Serpentine River smaller so that the water storage behind it would not drown the lake, and not damming the Huon River. This information had never been made public before and, unlike the rest of the HEC's evidence, was short on detail.[16]

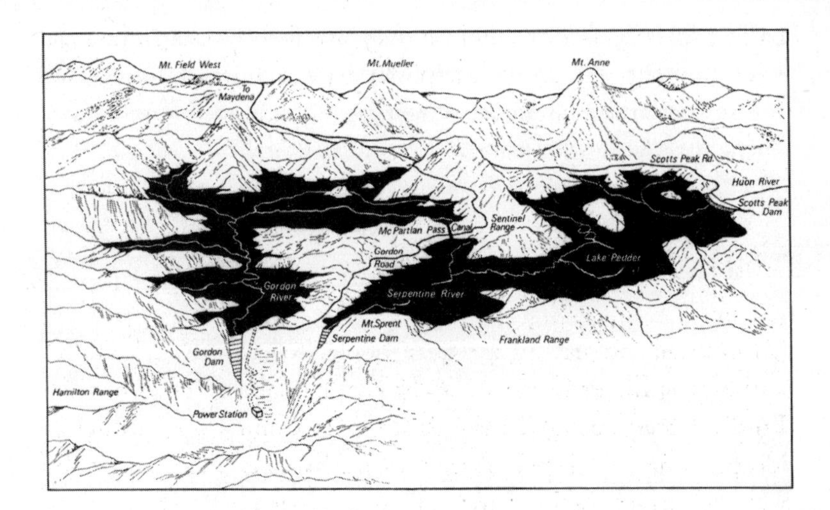

Figure 2.1 How Lake Pedder and Lake Gordon would look after flooding. (Source: *Pedder Papers* published by the Australian Conservation Foundation, 1972)

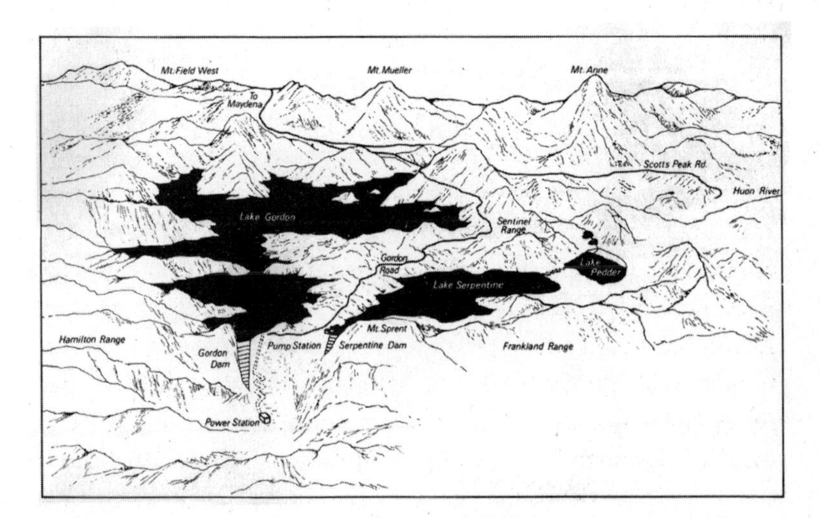

Figure 2.2 One of the alternatives that would have saved Lake Pedder, mentioned by the HEC at the Legislative Council inquiry. (Source: *Pedder Papers* published by the Australian Conservation Foundation, 1972)

The HEC's secrecy about the alternatives was arrogant; it showed no regard for public scrutiny of its proposals. Despite the new information, the inquiry recommended the HEC's preferred scheme go ahead, although it also said that the HEC should re-appraise its public relations policy, that a limit should be placed on the fluctuation of the lake level and that a thorough fauna and flora survey of the lake's environment should be done before it went under.[17]

The bill authorising the scheme took only two days to pass through the lower house then just another two to pass the upper house once the inquiry's report was handed down: record time for any piece of Tasmanian legislation. None of the upper house members who voted to destroy the lake had ever visited it. After their vote one member, Louis Shoobridge, did pay a visit. Upon his return he declared: '[after first glimpsing the lake from the air] I realised the enormity of what the state was going to do . . . like the man who comes face to face with Jesus Christ for the first time . . . it was no use landing on the lake floor . . . going home and saying it is still expendable . . . it was not to me any more expendable.'[18]

The conservation movement was shattered by the inquiry's findings and the passing of the legislation. It looked like the lake's fate was sealed and many gave up the fight. There was still widespread anger about the sacrifice of the lake but it seemed as though all avenues of appeal had been exhausted.

Once Reece had his legislation through parliament he was more impatient than ever to get the Lake Pedder hydro scheme under way and the controversy behind him. So the HEC began work on the dams that would flood Lake Pedder before it started work on the rest of the scheme. This ensured the lake was drowned a full seven years before the scheme was opened. Many questioned the haste but Reece and the HEC pushed on regardless; according to Reece, 'the sooner they fill it up, the better'. Knowing the lake's days were numbered, several scientific groups conducted field expeditions, discovering no less than thirteen new animal and four new plant species that were endemic to the area, all of them threatened by the lake's inundation.[19] Reece and the HEC remained unmoved.

A glimmer of hope for the environment movement came in May 1969 when Reece lost office to a new minority Liberal Party government led by Premier Angus Bethune, who relied on the support of a member, Kevin Lyons (son of Tasmania's only prime minister, Joe Lyons), the sole representative of the Centre Party in the state parliament. On occasion Lyons had been critical of the HEC, expressing the opinion that: 'We may yet come to the conclusion that we have given birth to a Frankenstein monster that will one day devour the body that established it'.[20] Bethune had himself referred to the HEC as a 'state within a state'.[21] But the environment movement's hopes were quickly dashed when the premier made it clear he was just as supportive of the Pedder scheme as Reece had been. Two months after being elected he went to the construction village for the Pedder scheme, Strathgordon, to trigger the first blast for the construction of one of the scheme's major water diversion tunnels.[22] Bethune reluctantly met with the conservation movement but gave it no joy. He told the Tasmanian Conservation Trust (a broadly based environment group formed in 1968): 'The issue can be simply stated. It is the value of Lake Pedder, which is a matter of opinion, set against the value of a power development, which can be assessed in real terms.'[23]

The second rising of the Lake Pedder campaign

By the start of 1971 the main dam that would flood Lake Pedder, the Serpentine Dam, was nearing completion. This focused the minds of the environment movement, which realised that unless a last-ditch stand was made the lake would quietly go under. The mood for another fight was enhanced by a major national political shift. The dam's completion coincided with a national growth of popular interest in social and environmental issues. In 1970 and

1971 hundreds of thousands of people took part in anti-Vietnam-war 'moratorium' marches around the country. In 1971 five members of the anti-conscription 'Save Our Sons' group were arrested. Also in 1971 there were major protests against the tour of a racially based South African rugby union team. Throughout 1969 and 1970 awareness of environmental issues was heightened by a popular campaign to save the Great Barrier Reef and by the 'Green Bans' movement, a 1971 union-led campaign to save remnant bushland along the Parramatta River in Sydney.

Global environmental awareness had been heightened by the 1962 publication of Rachel Carson's landmark anti-chemical tome, *Silent Spring*, the 1968 publication of Paul Ehrlich's *The Population Bomb* and by the imminent staging of the first United Nations environment summit in Stockholm in 1972. Specific awareness of the Lake Pedder issue was heightened by public showings of Lake Pedder slides by photographer Olegas Truchanas and by mainland campaign work undertaken by the Colong Committee and Milo Dunphy. Truchanas was a keen bushwalker and HEC employee who was committed to saving the lake. The flooding of the lake was therefore taking place at a time of heightened political and environmental awareness that made people eager to get involved. The upshot was that in Easter 1971 no less than 1500 people converged on the lake in a major demonstration of support and Louis Shoobridge, the MP who had visited the lake, called on the Bethune government to hold a referendum on the scheme. Bethune flatly refused. Out of this second rising came another organisation dedicated to saving the lake, the Lake Pedder Action Committee (LPAC), the most politicised and active of the three organisations that fought for it.

The LPAC took its campaign interstate by forming branches outside Tasmania and by pressuring the federal Liberal government of William McMahon to provide financial assistance to Tasmania to help save the lake. The issue was raised by the federal government at a premiers' conference in February 1972 but Bethune made it clear he was not interested in any federal rescue package.[24]

The LPAC was more than happy to tread the corridors of power and court the media. In Tasmania the organisation focused on a complete scrapping of the Pedder scheme although its Victorian branch took a softer line of campaigning for no damming of the Huon River and only allowing the Serpentine River to be dammed to below Lake Pedder. This was one of the alternatives identified by the Legislative Council inquiry. It would have reduced the inflow to the combined scheme by just 12 per cent.[25] The HEC tried to discredit this option by arguing that it would have withdrawn 20 per cent of the scheme's water, not 12 per cent. But it was using rubbery figures. It arrived at its 20 per cent figure by calculating the area that the damming of the Huon River represented, not the water volume it contributed.

Nineteen seventy-two was a salient year for the Lake Pedder campaign. In March the Bethune government collapsed after the relationship between the premier and Kevin Lyons broke down. State elections were called, whereupon it became clear an assortment of independent, concerned activists intended to stand for election on a save-the-lake platform. At a Hobart Town Hall meeting organised by the LPAC on 23 March a new political party, the United Tasmania Group (UTG), was formed, unifying this disparate grouping of independents. The UTG was the first environmental—or 'green'—party formed in the world, although its platform extended well beyond the environment. Its establishment came just two months before the creation of a similar party in New Zealand, the Values Party, which also came out of a campaign to stop a hydro electric scheme (on Lake Manapouri in the South Island). The motion passed at the town hall meeting read:

> In order that there is maximum usage of a unique political opportunity to save Lake Pedder, now an issue of national and global concern, and to implement a national, well researched conservation plan for the state of Tasmania, there be formed a single independent coalition of primarily conservation oriented candidates and their supporters.

The UTG was born, and although it had an expansive philosophy, the party's core agenda made it the political wing of the Pedder campaign (see appendix 2 for UTG's 'New Ethic'). The UTG was determined to make the lake an election issue even though the Labor and Liberal parties had a tacit agreement to keep it out of the election. During the campaign the LPAC and the UTG brought the lake to voters' attention through pamphlets, public meetings and newspaper advertisements. The large newspaper advertisements—in *The Australian, The Melbourne Herald* and *The Mercury* in Hobart—were the most prominent part of their campaign. They drew an immediate advertising response from the HEC, which was unused to having to justify itself to the public. One of the HEC's ads reproduced part of the LPAC's ads under the heading 'misrepresentation', then under the heading 'truth' it listed its responses.[26] By today's standards the advertisements (from both sides) were fairly wordy and unexciting but they marked the beginning of a new and enduring public phase in campaigning against hydro developments in Tasmania.

Although it had a broad policy base, a problem faced by the UTG—and by legions of Green parties after it—was convincing the media and the public that it was more than an environmental party. A leading figure in UTG, Richard Jones, said: 'the public and the media did us over. The only things they would publish would be the conservation component of any statement we put out—to demonstrate the validity of their claims that the UTG was a one issue mob. They never publicised anything that we would say on economics or alternatives—they wouldn't have a bar of it.'[27]

The UTG's best hope of getting someone elected lay with candidates Sir Alfred White and Ron Brown. White had been a state Labor Party minister in the 1940s and 1950s, and had recently been Tasmania's Agent-General in London. Brown had been a member of the state's upper house as well as head of the South West Committee. Both went close to being elected but ended up missing by a few hundred votes. Reece's Labor Party received a large vote, however, and was resoundingly returned to office. The

UTG was outraged by the blatantly political role the HEC played in the election and afterwards it unsuccessfully called for a Royal Commission into the HEC's role.

By mid 1972 the water backing up behind the Serpentine and Huon dams was starting to become a major reservoir, with the two impoundments inexorably moving towards each other. Lake Pedder did not have long to live and was due to be completely submerged by the middle of 1973. In June 1972 the lake's beach started to disappear. The LPAC ramped up its activity. There was an unsuccessful bid for a new upper house inquiry and a large number of petitions were presented to parliament, including one from 180 scientists who argued the lake should be saved for its endemic species. Another petition contained 17 000 signatures: the largest petition ever presented to state parliament.

One of the more intriguing developments in the campaign took place in July 1972 when the LPAC was anonymously tipped off that there might have been a major flaw in the legality of the legislation authorising the Pedder scheme. The group received legal advice that, in fact, the legislation was of dubious legality because the scheme had not been mentioned in the national park management plan for the area, as required under the new *National Parks and Wildlife Act* of 1970. The LPAC sought to initiate legal action as a result of this mistake but to do so required the permission of Reece's Attorney-General, Merv Everett. Although Everett was not obliged to give the LPAC legal standing, he was nonetheless inclined to do so in accordance with democratic fairness and due process, stating that: 'Legal rights are determined by judges of the courts, not by parliament. Once that principle is departed from I don't believe a truly democratic society exists.'[28] However, neither state cabinet nor the Labor Party caucus would back Everett so he was forced to temporarily stand down. Reece took over the Attorney-General's ministry for a short time and immediately announced he would not give the LPAC legal standing. He wasted no time in introducing 'doubts removal' legislation that closed off legal loopholes.[29] Reece wanted nothing to stand in his way.

The year also brought personal tragedy for the conservation movement: in 1972 three of its towering figures died. In January Olegas Truchanas died while canoeing on the Gordon River. As well as using his slides to publicise the Lake Pedder campaign, Truchanas had been instrumental in securing the reservation of a large stand of Huon pines on the Denison River, not far from Lake Pedder (see Chapter 4). He was a member of the Australian Conservation Foundation's council and had campaigned to save the Pieman River from hydro development. Then in September the movement was rocked by the mysterious deaths of Brenda Hean and Max Price. The two had set off in Price's Tiger Moth aircraft from Hobart to write 'Save Lake Pedder' messages in the skies above Canberra and to lobby federal politicians. Hean was a significant figure in the LPAC; Richard Jones called her 'the real driving force' behind the organisation.[30] The pair never arrived in Canberra and their plane was never found. Before they set off, the plane's hangar was broken into and the two received threatening telephone calls. The environment movement called for an inquiry but, incredibly, neither the government nor the police would agree to it. Journalist Chris Lewis in the Hobart *Saturday Evening Mercury* newspaper said:

> Who or what killed Max Price and Brenda Hean? The question will have to be answered ... the official silence, conflicting reports, dangerous allegations and general confusion cannot go on ... it is deadly serious when two people are missing in a plane that has allegedly been tampered with.[31]

Another major figure in the Lake Pedder campaign, Dr Bob Walker, called the deaths of Hean and Price 'one of the saddest parts of the campaign'.[32] The campaign for an inquiry into their deaths continues to this day. In 2007 a memorial service was held for them in Hobart on the thirty-fifth anniversary of their disappearance. Production of a film about their mysterious deaths also began the same year. The film makers offered a reward for conclusive evidence about how Hean and Price died.

The federal inquiry into Lake Pedder

One of the defining differences between the unsuccessful fight by the conservation movement to save Lake Pedder and the ultimately successful one to save the Franklin River was the preparedness or otherwise of each to appeal to the federal government. The Pedder campaigners appealed to the federal government fairly late in their fight whereas the Franklin campaigners began knocking on Canberra's doors early on. After the refusal of the Bethune state government in 1971 to hold a referendum on Lake Pedder's future and the re-election of Reece in 1972, it was obvious the federal government was the final hope for the lake. Jones said: '[after the 1972 state election] we moved on to the federal sphere . . . we knew there must be something that could be done there'.[33]

A federal election was held in December 1972 amidst a national mood for change after 23 years of conservative rule by the Liberal and Country parties. The LPAC hoped the federal Labor Party might save the lake. To make sure no one forgot about the issue during the election campaign, the LPAC parked a caravan outside Parliament House in Canberra and collared every politician it could find. In October, Reece claimed the federal opposition leader, Gough Whitlam, had assured him a new Labor government would not try to stop the Pedder scheme.[34] However, the environment movement's hopes were raised during the election campaign when Labor's federal environment spokesperson, Tom Uren, said: 'I commit myself to initiating a feasibility study of all the alternative proposals which could save Lake Pedder'.[35] In February 1972 he had also told federal parliament that the gates of the dam that flooded Lake Pedder should be reopened while a reassessment of the lake's flooding was undertaken.[36]

Whitlam won the election but instead of giving the environment ministry to Uren, it went to another member of Labor's left-wing faction, Dr Moss Cass. In January 1973 Reece had secret talks with Whitlam in an attempt to stop the inquiry promised by Uren. Reece came away thinking he had been successful. At the same time Cass made a visit to Lake Pedder and told the accompanying media that Lake Pedder was no longer just a Tasmanian issue.

Despite some early sympathetic words uttered about Lake Pedder, Whitlam ultimately decided he did not want to get involved in the issue, telling Cass to: 'stay out of Tassie'. Cass later said: 'It was clear he [Whitlam] was quite resolute in his mind that the whole thing was finished, the dam was now filling up and he was not going to do anything about changing that'.[37] However, in spite of Whitlam's opposition, Cass managed to convince federal cabinet to go ahead with the inquiry. In February 1973 the names of the people who were to conduct it were announced: Dr John Burton, a professor from the University of New England's natural resources department; Doug Hill, an engineer and hydrologist; Dr Bill Williams, a zoologist from Monash University; and Edward St John, a barrister and former Liberal Party member of parliament. The inquiry would be the last hope for preventing the lake from going under. A major motivation in Cass's push for the inquiry was that there had never been an environmental impact statement prepared for the scheme and he hoped the inquiry would go some way towards addressing that flaw.

In March the inquiry's terms of reference were announced. They seemed narrow, mainly concentrating on the events leading to the lake's flooding, and on that basis Reece initially agreed to cooperate. The inquiry visited Tasmania then held public hearings in April and May. Reece, however, became suspicious and after further communication with the federal government announced at the end of April that his government would no longer cooperate with the inquiry. The HEC soon followed suit (although it maintained limited involvement). Reece said of the inquiry: 'I'll tolerate no interference from Canberra or elsewhere ... that's the end of it'.[38]

In June, Cass released the inquiry's interim report. Its key recommendation was that a three- to five-year moratorium should be placed on the lake's flooding, during which consideration of alternatives could be revisited. The committee had been pressured by the federal environment department not to make the recommendation, but it stuck to its guns. Its thinking was influenced by the involvement of Doug Hill in an inquiry into some New Zealand hydro dams that demonstrated to him how biased hydro authorities could be.[39] The inquiry also recommended that the HEC be paid $8 million in compensation: the inquiry's estimate of the likely extra costs needed to make up for the loss of generating capacity during the moratorium.

Federal cabinet rejected the inquiry's recommendations. Cass summed up the mood: 'all the economic hardheads around the table were more concerned with development in those days and I did not get anywhere'.[40] When Cass reported the decision back to the Labor Party caucus, however, it backed the inquiry's recommendations and overturned the cabinet decision after viewing a Pedder slide show featuring photos by Olegas Truchanas. One of the Labor members who urged caucus to support the inquiry was Bill Hayden, leader of the party during the early part of the campaign to save the Franklin River.[41]

Whitlam was furious about the caucus rebellion but manoeuvred around it by sending a letter to Reece that simply notified him of caucus's decision. This allowed Reece to claim he never received a formal moratorium offer from Whitlam. Cass explained:

> [the Tasmanian government] never heard from the [federal] government. I thought that Gough had never told Reece at all but I subsequently learnt that he sent a letter simply saying caucus passed the following resolution, and gave the text of the motion. Full stop. And Reece very cleverly said—and maybe

> they hatched it up between them—'I will wait till I hear the government's decision'. Now the government's decision had to be conveyed by Whitlam. Whitlam never said anything about the federal government offering financial support. It was just a caucus decision.[42]

Whitlam's dead hand killed off any remaining hope of saving the lake and by early 1974 the original Lake Pedder was well and truly flooded. Whitlam is often justifiably hailed as a great Australian reformer but his record on Lake Pedder was poor, to say the least.

Some of the most poignant text in the inquiry's final report was written by Edward St John, who had had an earlier involvement in the anti-apartheid movement. In a supplementary statement to the report he said: 'I am quite certain that the decision to flood Pedder was a great mistake'.[43] He noted that 'the Huon waters [that flooded the lake] were not necessary to the scheme; they contributed only twelve per cent of the average annual water flow',[44] and expressed the hope that 'it should still be possible, over a period of years, to see Lake Pedder restored'.[45] He later uttered some of the most memorable words from the campaign: '[there is] an opportunity to repent . . . if not we ourselves, our children will undo what we so foolishly have done'.[46]

The scheme was finally opened in 1979, at which the Tasmanian Wilderness Society staged a small protest. Reece said it made no difference to man, woman or child that the lake was drowned. In the end the cost of the combined scheme doubled to $190 million, making the extra $11 million the HEC claimed it would have cost to save Lake Pedder seem insignificant. (Even the $11 million costing was questionable, including, as it did, the costs of pulling down major dam walls and labour-intensive environmental restoration.)[47]

In subsequent decades there were several postscripts to the controversy. In 1986, the leader of the Tasmanian parliamentary Greens, Bob Brown, released a book of Lake Pedder photographs that sold out within months. In 1990 the minority state Labor government—kept in power by the Greens—asked the HEC to investigate the feasibility of restoring the lake but, predictably, the HEC advised against it. In 1994 Brown and others set up Pedder 2000, a lobby group dedicated to keeping alive the idea of draining the Pedder dam. At the organisation's launch Brown declared: 'It seems a small price to pay to show that our generation is not only capable of massively changing the environment, when it needs to, but can also accept the challenge of restoring it to its former beauty. It also shows that our society can still give the thumbs up to those with great dreams.'[48] Pedder 2000 is still active today, arguing that Tasmania could lead the world in restoration ecology. In 1995 a federal government House of Representatives committee also investigated the feasibility of restoring the lake but again advised against it. The chairman of the committee, John Langmore, said: 'While the proposal would be technically feasible and would, if implemented, enhance the World Heritage values of the area, there is some doubt that the environmental benefits would offset the costs'.[49] Despite the adverse conclusion, the committee's argument that the draining of the Pedder dam was feasible was a significant win for the conservation movement. In 1995 a major symposium was also held at the University of Tasmania on the feasibility of restoring the lake.

Despite the loss, the flooding of Lake Pedder left some positive legacies for Tasmania and Australia. One was the creation of both the National Parks and Wildlife Service in Tasmania and the first major national park in south-west Tasmania. Another was the establishment by the Whitlam government of a National Estate heritage inventory via the *Australian Heritage Commission Act* (which it passed in 1975 in response to various environmental issues including Lake Pedder). Yet another was the Whitlam government's signing of the World Heritage Convention in 1974, which played a pivotal part in saving the Franklin River. And a further significant legacy was the unprecedented public scrutiny the Pedder campaign exposed the HEC to. The commission had been used to never having to defend its development proposals but after the Pedder fight it had to make a public case for any new scheme.

The flooding of the lake also left a number of specific positive legacies for the environment movement. Probably the most significant one was the part played by the Pedder campaign in forever changing conservation battles in Australia. They went from polite, discreet meetings with officials to major campaigns to win the hearts and minds of voters. More than other environmental campaigns in Australia, the fights of the late 1960s and early 1970s to save the Great Barrier Reef and Lake Pedder were responsible for bringing about this seismic change. The bureaucratic style of the South West Committee was out and the louder, more politicised style of the Lake Pedder Action Committee was in. Part of the sea change was a realisation that a major part of modern environmental campaigning had to involve significant use of images that could bring the grandeur of wilderness into peoples' living rooms. As Bob Brown said: 'The difference between the Lake Pedder campaign . . . and the Franklin campaign . . . was that we moved from black-and-white television images to colour television images and everyone could see it'.[50] After Lake Pedder the environment movement made much greater use of imagery in its campaigns.

The switch to a more public and assertive environmental campaign style ended up having major consequences for Australia's major national conservation organisation: the Australian Conservation Foundation (ACF). During the Pedder fight, many felt both it and the Tasmanian Conservation Trust had not been sufficiently involved and had taken a fairly 'hands off' approach. Their disinterest allowed Premier Bethune to say: 'prominent and responsible conservationists are not interested in the issue'.[51] The tension came to a head at the 1973 annual general meeting of the ACF, chaired by Prince Philip, where a number of Pedder activists successfully led a charge to remove the group's conservative hierarchy.

The Pedder campaign had negative legacies too; most obviously, the ultimate failure of the campaign's main aim to save the lake. An additional negative legacy was that it left the environment movement with no energy or morale to fight the second major scheme the HEC planned for western Tasmania: the Pieman scheme. The Pieman scheme was approved by the Tasmanian parliament in 1971 while the movement was distracted by the Pedder campaign. Work began on it in 1973. Although it would not affect the wilderness areas of western Tasmania to the same extent that the Pedder scheme did, it involved the diversion of the Murchison River into the Mackintosh River and the construction of a large dam on the Murchison River. The waters from the dam reached back inside the then western boundary of the Cradle Mountain–Lake St Clair National Park. The flooding of part of the national park even triggered the Pieman scheme's own 'doubts removal' legislation and change to the national park's boundary just as there had been for the Pedder scheme. By rights, it was a development the conservation movement should have fought but the focus on Lake Pedder allowed the Pieman scheme to sneak under the radar. One person who did campaign for the Pieman, however, was Olegas Truchanas. He took many compelling photos and lobbied the federal government to save the Pieman, but to no avail. He mounted a one-man campaign in its defence and was disappointed the conservation movement did not devote more energy to it.

A further negative legacy of the Pedder fight was the subsequent polarisation of opinion over all major conservation issues in Tasmania. Apart from a brief period in 1990, when the short-lived Salamanca Agreement (see Chapter 5) tried to find common ground in the state's forests debate, consensus has been conspicuously absent in Tasmania's wilderness battles since Lake Pedder. The polarisation was much worse during the Franklin River debate but it started with the Pedder controversy and Tasmania has never recovered from it. Some commentators argue the Vietnam War forever polarised the United States—the same can be said of Lake

Cartoon by Ron Tandberg that appeared in *The Age* in 1982 commenting on hydro development in Tasmania.

Pedder's effect on Tasmania. In many ways, each new wilderness battle in the state is a repeat of the Lake Pedder experience.

The loss of the lake was a bitter blow to all who fought for it, many of whom could not face being involved in wilderness battles again. Max Angus ruefully declared that: 'we were trodden underfoot because we were idealistic'.[52] Bob Walker summed up the anger and frustration of campaigners: 'They did it [the flooding] in totally engineering terms and mankind will condemn them for it . . . it was put up to the public as a fait accompli, no consultation . . . who can imagine what people will think of it all in one hundred years time?'.[53]

The flooding of the lake still invokes strong emotions and will never be forgotten.

3

The Franklin and beyond

The failure of the campaign to save Lake Pedder left deep wounds in the Tasmanian conservation movement and many activists retired from the fight against the Hydro Electric Commission (HEC). Those who remained knew that the Franklin River was the next target. Once Eric Reece had announced in 1962 that three separate hydro schemes were planned for the Gordon River (into which the Franklin flowed) the conservation movement knew the Franklin was vulnerable, although little could be gleaned about how advanced the HEC's plans were or when the new battle would begin.

Having been exposed for its lack of alternatives to the Pedder scheme, the HEC made a deliberate show of investigating alternative dam sites to the Franklin dam. But it was mainly interested in one site. In 1971 it built an access track into the middle reaches of the river to aid investigations and by early 1977 had settled on a point 400 metres downstream of the junction between the Franklin and Gordon rivers as its preferred site. In June that year the HEC formally announced it intended flooding the Franklin. The first stage of the scheme would include a dam on the Gordon that would flood the lower Franklin. A second stage was planned for the middle reaches of the Franklin that would hold back water from its upper and middle sections as well as water diverted from the King River near Queenstown, in Tasmania's west coast region. By the time Lake Pedder was being flooded the HEC had already spent $1.8 million investigating the Franklin scheme, with costs reaching nearly $4 million by 1977.[1]

The growing campaign against the Franklin dam

Interest in the Franklin from the conservation movement stepped up after Launceston forester Paul Smith invited a local doctor, Bob Brown, on a trip down the river in 1976. Their trip was one of the first to successfully navigate the treacherous river. (In 1958–59 John Hawkins, Henry Crocker, Trevor Newland and John Dean had managed to get down it in canoes, on their third attempt, and in 1970–71 another group managed to get down on a homemade raft that came apart.) Brown was deeply moved by the journey, declaring afterwards: 'The process of thirty years which had made me a mystified and detached observer of the universe was reversed and I fused into the inexplicable mystery of nature'.[2]

The experience seared onto Brown's mind the importance of saving the Franklin and he ended up making the river journey seven times between 1976 and 1981. The Tasmanian Wilderness Society (TWS) was formed a few months after Brown's first trip by sixteen people who gathered at his home at Liffey, in the north of the state. TWS grew out of the South West Action Committee, the group that had carried the flame of protecting Tasmania's wilderness since the end of the Pedder campaign. Later that year Brown held a lone fasting vigil on top of Hobart's Mount Wellington, in protest against the visit by a US nuclear-powered warship, the *Enterprise*. Despite an inclination to get further involved in the anti-nuclear issue, Brown took over the directorship of TWS in early 1979 after it was vacated by Norm Sanders who, like him, would end up becoming a conservation-based member of the Tasmanian parliament.[3] By then the Franklin campaign was well and truly TWS's top priority. Brown quickly warmed to the theme of telling people how threatened the river was. In early 1980 he told *New Idea* magazine: 'within a few years, the wilderness areas of the world—and there are only a few left—will be gone. We have a chance to save this one in Tasmania, not so much for this generation but for the future, and not just for Australians but for mankind.'[4]

Brown and TWS were keen to use visual material as much as they could in the Franklin campaign so Smith took movie footage of the river on a second trip in 1977. After showing it to state parliamentarians, TWS bought space for the footage on commercial television in 1979.[5]

In May 1977 vague details were released about the Franklin scheme then, finally, in October 1979 the HEC announced the specifics. It would be a $366 million, 296-megawatt-capacity scheme which the HEC estimated would add an extra 22 per cent to the state's power-generating capacity (once the scheme's second stage was constructed) and would be completed by the early 1990s. At last the truth was out. TWS swang into action, organising a rally in Hobart and a packed public meeting at the city's town hall. It also organised 40 000 signatures for a 'Save the Rivers' petition and published a 32-page colour booklet of images of the Franklin and Gordon rivers that was launched by TWS's patron, the violinist Yehudi Menuhin.

TWS was prepared to try anything to raise awareness of the Franklin. Brown said: 'We grabbed ideas from wherever we could. We looked at the way other people who sell cheese and paper tissues, how they do it, and thought if that sells an idea then how much more important that that be grafted by us into saving a wilderness.'[6] Two other innovative ideas he and TWS used were to commission an opinion poll that showed Tasmanians overwhelmingly wanted the Franklin saved and to dress in suits when addressing the media or politicians. This deliberate attempt at mainstream appeal was new for the environment movement. Brown said: 'if you have a television camera aimed at you, a lot of the viewers at home would not be able to communicate if people looked strangely dressed'.[7] TWS's new high-profile, broad-based style seemed to resonate with the public: its membership grew from 200 to 1000 in just twelve months. By mid 1980 it had reached the hitherto undreamt of level of 2000 members.

Despite the unprecedented scrutiny the HEC was exposed to during the Pedder campaign, Brown and TWS were pitted against an organisation no more accountable nor open to probity than it had been during the Pedder days. Some in the environment movement even referred to it as 'the Electric Kremlin'. As if to confirm its insularity, in 1980 the head of the HEC, Russell Ashton, declared: 'If the parliament tries to work through popular decisions we are doomed in this state and doomed everywhere'.

Lowe's compromise scheme

Once the HEC released its Franklin scheme proposal, events moved quickly. In May 1980 a committee that Labor Premier Doug Lowe had established to provide energy advice (independent of the HEC) reported that the HEC had overestimated future electricity demand. It recommended that instead of building the Franklin scheme, a less damaging dam further upstream on the Gordon, above its junction with the Olga River, should be constructed.[8] The Olga scheme had originally been put up by the HEC as a dummy alternative: it did not think it would be taken seriously.[9] But the scheme gave Lowe a way out of the increasingly difficult issue. Lowe had a difficult time convincing his cabinet about the wisdom of backing the Olga compromise but got a majority behind his preferred solution, largely thanks to a supportive minister, Michael Barnard.[10] Had a prominent pro-Franklin dam minister, Deputy Premier Neil Batt, not been overseas when cabinet debated the issue Lowe might have lost. Lowe allowed Bob Brown to briefly address the cabinet meeting. Before he sat down, however, one minister told Brown: 'this is the last time you'll see the inside of this room!'

When announcing his government's backing for the Gordon-above-Olga scheme, Lowe said his government would create a new Wild Rivers national park that would preserve the Franklin's catchment and surrounding wilderness. The Wild Rivers national park idea had been actively promoted by Lowe's national parks minister, Andrew Lohrey, and the National Parks and Wildlife Service in defiance of the HEC. Lohrey took the idea to cabinet several times before it was accepted. Many wilderness advocates were initially elated at the news of the Gordon-above-Olga compromise. It took them some months to realise the Olga option was just a case of switching one invasive, damaging scheme for another. It would still take hydro development further into the south-west, much like the Franklin scheme would have. Lowe, however, saw the scheme as a significant rebuttal of the HEC. When introducing the legislation authorising the Olga scheme he said:

> The HEC is an engineering organisation . . . it is not a socio-economic planning body. Previous governments may have been satisfied with a cursory perusal of the HEC's recommendations, followed by an automatic stamp of approval. This is not my style.[11]

Lowe was hoping for support for the Gordon-above-Olga scheme from the opposition Liberal Party, as well as from TWS, but on both counts he was disappointed. The deputy leader of the Liberal Party, Robin Gray, successfully urged his party to back the Franklin scheme. TWS was relieved that Lowe's compromise would save the Franklin but came to see it as a pyrrhic victory. Brown congratulated the government for standing up to the HEC but said TWS would campaign against the Gordon-above-Olga scheme nonetheless (partly because, like the Franklin scheme, it would flood the picturesque Gordon Splits Gorge on the Gordon River).[12] Brown told the media: 'We are disappointed the government has chosen the Gordon-above-Olga scheme but for now we're celebrating . . . the Franklin has been saved . . . it is an historic decision because it is the first time a Tasmanian government has overturned a major development project in favour of a national park.'[13]

The lukewarm reaction to his compromise left Lowe vulnerable. Many in his party supported the Franklin scheme and the reaction

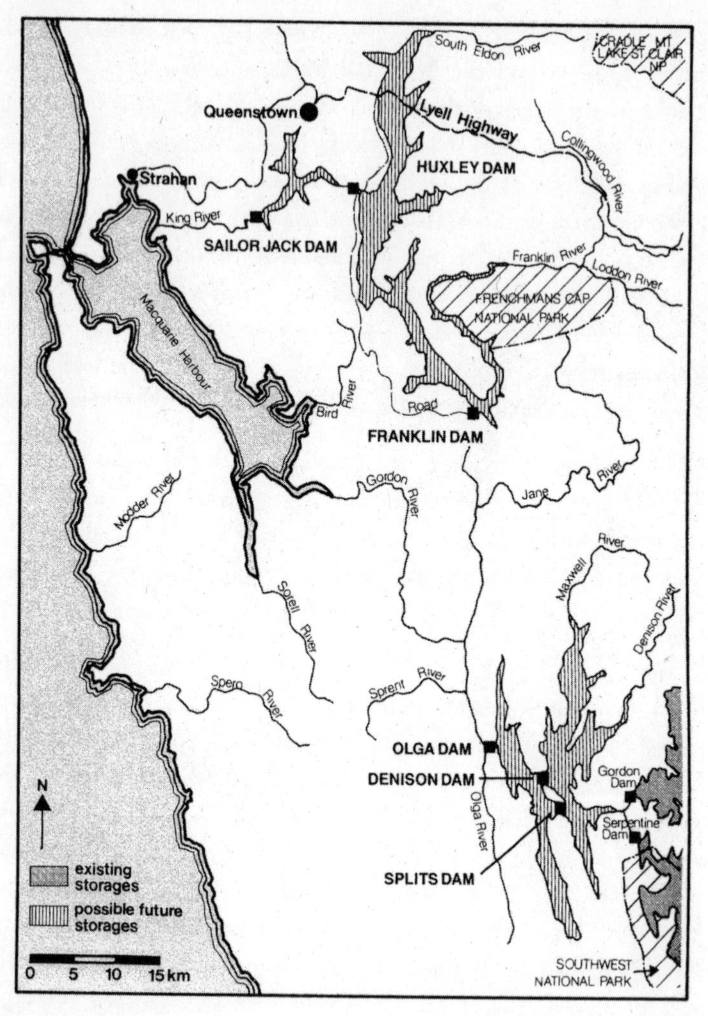

Figure 3.1 Doug Lowe's preferred Gordon-above-Olga hydro scheme. (Source: Hydro Electric Commission)

to the Gordon-above-Olga option was not favourable enough to quell their disquiet.[14] Lowe was not the only one left vulnerable by the Gordon-above-Olga decision: interest in TWS's campaign to save the Franklin waned after it was announced and fewer people volunteered their support to the organisation.[15]

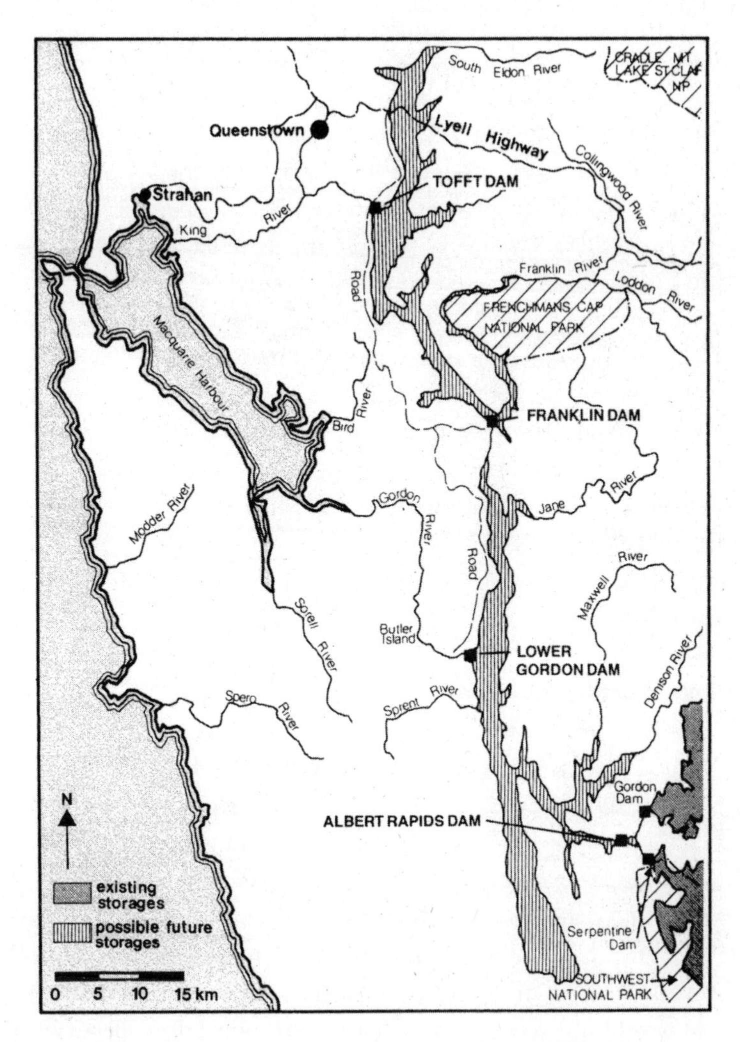

Figure 3.2 The HEC's preferred Franklin River scheme. (Source: Hydro Electric Commission)

Around the same time, the cultural significance of the Franklin River area was enhanced by the discovery of a cave on its lower reaches that contained evidence of the southernmost existence of humans during the last ice age. To put pressure on the Prime

Minister, Malcolm Fraser, it became known as 'Fraser Cave'. The cave was eventually named Kutakina Cave, and carbon-dating of its remains revealed them to be 24 000 years old.

Soon after Lowe announced his backing for the Olga scheme, the upper house of Tasmania's parliament set up an inquiry into the competing schemes, much as it had with the Lake Pedder scheme. Lowe had originally wanted an inquiry conducted by both houses of parliament but failed to secure crucial support from the opposition (which wanted a non-parliamentary inquiry). The upper house went ahead with its own.[16] In a brazen move that was probably outside its legislated powers, the HEC made a submission to the inquiry that was critical of the Gordon-above-Olga decision (in a repeat of the inflexibility it had shown towards alternatives to flooding Lake Pedder).

In November, the legislation authorising the Gordon-above-Olga scheme passed Tasmania's lower house. But the upper house's inquiry backed the Franklin scheme, not Lowe's compromise scheme. It refused to pass the Olga legislation, instead substituting the name of the Franklin scheme into the legislation. This move was of dubious constitutionality and was based on advice obtained by the Tasmanian Chamber of Industries from a lawyer retained by the HEC.[17] Many in TWS, including Brown, thought the Gordon-above-Olga scheme would be unstoppable but the upper house's arrogance gave them a ray of hope: it still might be possible to have no scheme at all. The upper house's refusal to pass the Olga legislation was crucial to the Franklin issue: without it the campaign would have stopped in its tracks once the compromise scheme went through. The upper house unwittingly handed the conservation movement a lifeline.

With the upper house, TWS, the Liberal opposition and many in his own party refusing to back the Gordon-above-Olga option, Lowe became increasingly isolated. In July 1981 he narrowly survived a leadership challenge from rival Harry Holgate, who supported the Franklin scheme. In an attempt to boost support for his compromise, earlier in 1981 Lowe had announced that the new Wild Rivers National Park would be nominated, along with the Southwest and Cradle Mountain–Lake St Clair national parks, for World Heritage listing. He was hoping this would improve the environmental credentials of his middle path. Prime Minister Malcolm Fraser supported Lowe's World Heritage nomination. The two first talked about the nomination in 1980 and Fraser made it clear to Lowe from the start that the nomination would have his wholehearted support.[18]

To try to give his compromise option additional credibility, in September 1981 Lowe got the support of his Labor parliamentary colleagues for a statewide referendum on the scheme, but events soon got ahead of him. The Franklin issue was playing out with the same heightened passions that infused the Lake Pedder controversy. Lowe's support was crumbling and in October Holgate managed to topple him. Holgate then abandoned the government's support for the Gordon-above-Olga scheme, switching it to the Franklin scheme. Holgate's coup, and his switch of support, revived interest in TWS's campaign to save the Franklin. Just before Holgate took power, however, Lowe signed the proposed World Heritage listing and it was despatched to the federal government (partly thanks to a diligent member of the National Parks and Wildlife Service) and Fraser sent it to the World Heritage Bureau in Paris in November.

This pre-emptive move was crucial to saving the Franklin. Lowe and his cabinet originally made the Wild Rivers National Park and its World Heritage listing part of a package and they felt it would have been disingenuous for their government to proceed with that part of the package before the Legislative Council passed the Gordon-above-Olga legislation. But after it became clear in April 1981 that the upper house would never pass the Olga legislation, Lowe changed his mind and proceeded with the national park proposal and World Heritage nomination without the upper house's support.[19] Had he not had this change of heart, the Franklin River never would have been saved. World Heritage listing made it possible for the federal government to intervene in the issue.

Lowe came to be derided by both sides of the Franklin debate for his middle path approach but he deserves more credit than he ever got for unilaterally pursuing the World Heritage nomination. He made it possible for the federal government to save the river. After his loss to Holgate, Lowe dramatically quit the Labor Party, along with fellow Labor renegade Mary Willey, and sat on the parliamentary crossbenches. Together with Democrat Norm Sanders, the three had the balance of power in the lower house with Holgate's government reduced to seventeen out of 35 seats. Despite Lowe's leaving the Labor Party, however, the referendum he promoted went ahead in December 1981.

The dams referendum

TWS was not prepared for the huge campaign the referendum necessitated—it lacked both funds and energy (and was even on the brink of insolvency)—but thanks to the Herculean efforts of a few individuals, especially Peter and Shirley Storey, it mounted an effective statewide campaign, involving one hundred volunteers who knocked on virtually every household door in the state. TWS's campaign was aided by the Holgate government's arrogant decision not to include a no-hydro-scheme option on the ballot paper. Holgate had enormous pressure put on him to withdraw a no dams option by union leader Leo Brown, who was president of the Tasmanian Labor Party.[20] Once again, political overreach came to TWS's rescue. In response, TWS urged Tasmanians to cast an invalid vote by writing 'No Dams' on their ballot papers. This made the referendum a democracy issue as well as a conservation one. It was an audacious move by TWS that would not necessarily pay off: it was asking a lot of ordinary citizens to invalidate their vote.

TWS need not have worried. The initial count gave the Franklin scheme 53 per cent, the Gordon-above-Olga scheme just 9 per cent and informal votes 38 per cent.[21] Following legal advice to TWS that ticks and crosses could not be counted as valid votes, support for the Franklin scheme fell to 47 per cent, the Gordon-above-Olga scheme to 8 per cent and the informal vote rose to 45 per cent.[22] With no clear mandate for either of the two schemes, TWS scored a major victory that made people around the country sit up and take note of the Franklin campaign for the first time. The referendum gave TWS a much higher profile and its membership started soaring, again, as a result. Holgate came out a clear loser and, lacking both public and parliamentary support, put off resuming parliament until March the following year. He knew that as soon as parliament resumed his government would fall and his life as Premier would probably be over.

When the Tasmanian parliament resumed there were no surprises when Holgate lost a no confidence vote—with Sanders, Lowe and Willey voting with the Liberals to bring him down—and elections were called for May 1982. Sanders stood for his seat again but Bob Brown also ran for the same seat as an independent (in the seat of Denison, which had seven members elected from it under the state's multi-member 'Hare–Clarke' electoral system). The Liberals, who by then were led by Gray, won a landslide majority and Holgate was trounced. Sanders was re-elected but Brown narrowly missed out. For Brown it was an agonising election: he knew a lot of major power groups were against the fight to save the Franklin, including the Labor Party, the Liberal Party, the media and the unions. The election was therefore an arduous experience for him and the conservation movement in general. After Gray's win, a prominent member of the Legislative Council, Harry Braid, confidently declared: 'no power on this earth will stop the dam now'.

Once elected, Gray wasted no time introducing legislation authorising the Franklin scheme and by June 1982 it had been approved by parliament. The HEC began work on an access road into the area the next month. Gray asked Prime Minister Malcolm Fraser to withdraw Lowe's World Heritage nomination but Fraser refused. Within a few weeks of the state election, Brown went to see Fraser in Canberra only to be told the Franklin was a state issue and the federal government would keep out of it.[23] Fraser was one of the founding members of the Australian Conservation Foundation, and therefore had some conservation sympathies, but he was not prepared to go beyond the World Heritage nomination. He had intervened in the campaign to save Fraser Island, in Queensland, but would not do the same for the Franklin. A defiant Brown declared: 'we'll fight to the last bucket of cement'.[24]

As well as lobbying politicians, TWS also pursued a legal strategy, taking a case to the High Court seeking to deny all-important loan funds to the Tasmanian government for the dam's construction. The case was dismissed, however, and TWS was forced to look at other strategies. A small victory for TWS in the case was its ability to convince the court that it had legal standing in the issue. The Tasmanian Conservation Trust had courageously, and unsuccessfully, fought for that right during the second half of the 1970s (see Chapter 8, 'The Precipitous Bluff fight').

In a federal by-election in March 1982, 12 per cent of voters in the seat of Lowe had written 'No Dams' across their ballot papers, giving TWS confidence that the Franklin issue had national political appeal. Fortunately for TWS, in July the Labor Party held its biennial national conference in Canberra and the organisation saw a chance to get Labor behind its Franklin cause. The party conference made a decision to allow uranium mining to proceed in Australia so things did not look good for TWS's Franklin motion. However, thanks to some sympathetic chairing, and filibustering, by New South Wales Premier Neville Wran the conference ended up passing a motion opposing the Franklin scheme.[25] The result was close, the motion being carried by just three votes (then federal leader of the Parliamentary Labor Party, Bill Hayden, opposed it). Brown had met Wran's wife, Jill, at an earlier function where she expressed a keen interest in showing her husband material on the Franklin. Early in the campaign, TWS also secured the enthusiastic backing of the Democrats, whose leader, Don Chipp, had rafted down the river with TWS campaigners in 1981. Chipp went on to speak at many Franklin rallies and always counted (perhaps overgenerously) the Franklin campaign as one of the great successes of his party.

The economic arguments against the dam

Apart from superior organisational skills and better use of the media, another feature that distinguished the campaign to save the Franklin River from the fight to save Lake Pedder was the preparedness of environmental campaigners to comprehensively tackle the economic case against hydro development. The economic arguments covered several fronts including employment and the cost to the state.

TWS argued hydro development was having a negative impact on the state's employment levels. In the mid 1970s Tasmania's level of unemployment had been equal to the national average but by 1982 it was 40 per cent higher[26] and has remained at least 20 per cent higher ever since. The rise in the state's unemployment rate was partly caused by a 7 per cent fall in the employment of the state's major industrial users between 1975 and 1985 (with cuts continuing to this day).[27] This came despite a doubling of the power consumption of the state's major industrial users—who used two-thirds of the state's electricity—during the 1970s.

Mainly as a result of its hydro mania, by the 1980s Tasmania's level of per capita state government debt was the highest in Australia and was a third higher than that of the next highest state.[28] TWS argued the Franklin dam would end up adding the best part of another 1 billion dollars to the state's debt and would make its debt levels unsustainable. The power prices being offered to major industrial users were a secret but were thought to be less than one-tenth of those paid by household consumers. This was confirmed when details of bulk power contracts were leaked to TWS in 1983. Hugh Saddler, a leading national energy analyst, argued the contracts showed the state was paying a subsidy to its bulk electricity consumers of about $60 million per year.[29] Given the state's rising unemployment and debt levels, TWS argued this was far too high a price to pay.

Proposals about what type of economic direction Tasmania should pursue instead of hydro industrialisation made up a third front in TWS's economic case. TWS said that instead of investing in energy consumption, Tasmania should invest in energy conservation and renewable energy generation. It also said the state should support labour-intensive industries while trading on its 'clean, green' image by enhancing its tourist industry. A 1980 study commissioned by the Tasmanian Conservation Trust from two University of Tasmania academics found that for the same government outlay, the Tasmanian government could create more jobs in energy conservation, renewable energy generation and labour-intensive business creation than it could by spending money on more dams.[30] A 1981 report commissioned by a small business group, the Business Association for Economical Power, written by economist Shann Turnbull, also concluded that energy conservation would be more economically advantageous for the state.[31] TWS built a convincing economic case that gave its cause enhanced credibility.

The Franklin blockade

By mid 1982, as well as fighting on a federal front, TWS was secretly planning a blockade of the Franklin dam construction site on the Gordon River. Like its referendum campaign, this was a bold strategy. The blockade set TWS apart from the conservative conservation groups that had fought the early days of the Lake Pedder campaign. TWS could not be certain such an audacious move would be successful but the group felt it was a necessary strategy to keep the Franklin issue alive. Major direct action protests at that time were rare, and had only been conspicuously used in Australia once before, in 1979 as part of the ultimately successful fight to save the Terania Creek rainforests in New South Wales. TWS decided to use the law-defying tactic even though some around the organisation, such as Sanders, feared it could get violent. To allay such fears, TWS emphasised the pacifist nature of the blockade and even brought in overseas specialists in non-violent protest. TWS insisted that all blockade participants be trained in non-violent methods.

In August at Hobart's prestigious Hadleys Hotel, and packing the audience with its most conservative-looking members, TWS announced to the outside world its intention to hold the blockade. In response, the following month the Gray government revoked the part of the Wild Rivers National Park that took in the dam construction site and vested it in the hands of the HEC. It also passed regulations that imposed a $100 fine, or six months' gaol, for anyone caught protesting on the HEC's land. TWS hit back by holding its own 'redeclaration' of the area. Gray then outraged many by saying of the Franklin: 'For eleven months of the year the Franklin River is nothing but a brown ditch, leech ridden, unattractive to the majority of people'.[32]

In October and November 1982 major demonstrations, attracting 4000 people in Sydney and 15 000 people in Melbourne, were the first significant manifestations of widespread national support for saving the Franklin. The demonstrations came on the back of a successful August national tour by Brown that did much to raise national awareness. Brown and TWS were determined to give the campaign a national profile as early as possible and not to leave it until it was too late as had happened with the Lake Pedder campaign. A TWS meeting in November decided the blockade would start on 14 December unless the federal government intervened beforehand.[33] Soon after, acting Prime Minister Doug Anthony announced the federal government would definitely not be intervening so the blockade started as planned, resulting in the arrest of 53 people that day.[34] On the same day the 'Western Tasmania Wilderness National Parks' were inscribed on the World Heritage list despite last-minute attempts by the Tasmanian government to stop it. Gray said the listing was 'a stupid mistake based on a total lack of understanding'.[35] The area joined other iconic sites such as the Grand Canyon and the pyramids of Egypt as a place of international cultural and/or environmental significance. It satisfied all four of the natural criteria for World Heritage listing as well as three of the seven cultural ones: a record number for an Australian World Heritage site. The Western Tasmanian Wilderness National Parks was the fourth site listed by Australia. (The Great Barrier Reef, Kakadu National Park and Willandra Lakes were listed in 1981.) Today, Australia has seventeen listed sites. The World Heritage Convention is administered by the United Nations. In 1974 Australia became one of the first countries to ratify it. Currently over 830 sites from around the world are listed.

Ten days before the World Heritage listing, confidence in the electoral appeal of the Franklin issue received a boost when 40 per cent of voters in the Victorian seat of Flinders wrote 'No Dams' across their ballot paper in a federal by-election.[36] On 16 December 1982, Brown was arrested at the blockade, refused bail (as many blockaders did) and spent Christmas in gaol.

No other civil disobedience action has surpassed the Franklin blockade: it was the largest such action ever held in Australia. It pitted machinery against people in the middle of Tasmania's wilderness and created imagery that television and newspapers could not get enough of, particularly during the 'slow news' summer period. Huge bulldozers were towed on barges to the dam construction site on the Gordon River. As they drew near to their destination a long line of blockaders in rubber rafts attempted to stop them. Once at their destination the machinery smashed through trees and undergrowth only to be confronted by more lines of protesters. Police were close behind and quickly arrested the protesters. Relations between the two groups were generally amicable, with some police even secretly carrying anti-dam stickers inside their hats.

At first, TWS thought only modest numbers would take part in the blockade but the protest far surpassed expectations. Key organiser Lillith Waud said: 'We just had no comprehension of how enormous it would end up being, at the beginning. I have got a copy of a letter in which I wrote ... if we could get 200 we will be lucky.'[37] In the end 2613 people registered for the blockade at TWS's Strahan information centre—more than ten times Waud's original estimate—of whom 1272 were arrested.[38] The protest drew some high-profile visitors, including British environmentalist David Bellamy and electronics entrepreneur Dick Smith. Robin Gray was not impressed; he called the blockaders 'totally irresponsible ... extremists [and] fanatics'.[39]

The popularity of the blockade put an enormous strain on both TWS and the blockade facilities. Volunteer resources were stretched to the limit as TWS attempted to cope with the demands of having its headquarters in Hobart, on the eastern side of the state, while thousands took part in the blockade near Strahan on the western side. Communication was often poor and misunderstandings were frequent. Within the blockade organisation tensions also developed between the 'up river camp' on the Gordon River, where most of the direct action took place, and the blockade organisers based in Strahan.

On top of all this, Strahan residents were divided on the issue. Many were fiercely opposed to the blockade and were not afraid to express their hostility by shouting, spitting and making life as uncomfortable for the blockaders as possible. There was even a sticker that read 'Doze in a Greenie: Fertilise the South-West'. One night a group of local youths took to Brown with a tyre lever.

For the organisers it was an extraordinarily tense and exhausting time. Waud said: 'it was the most stressful period of my life ... I stopped seeing myself almost as a human being'.[40] It could all have turned violent, and might have ended up being a major public relations disaster for TWS, had it not been for their thorough non-violent training, which held the blockade together. Also making a major contribution to the success of the blockade was TWS member Cathie Plowman, who organised the action with Waud. Plowman said: 'It was an excellent blockade. We got what we wanted: worldwide attention and support. We had extensive and generally supportive media coverage right across Australia, not just for days but for weeks on end.'[41]

The blockade was successful not because it significantly slowed dam construction work but because it drew national attention to the threat facing the Franklin. Bob Brown summed up its coverage-garnering importance by saying: 'The Franklin blockade did not stop one bulldozer. But it did stop the dam. It allowed the beauty of the river to speak through TV screens in millions of living rooms to every Australian. And it elevated the environment to national thinking.'[42]

The blockaders deservedly got most of the plaudits for making the direct action the success it was but the person who allowed it to happen was Strahan tourist boat operator Reg Morrison. Originally TWS had planned a blockade away from the dam site at a location accessible by road but Morrison's intervention made a much more ambitious direct action possible. Morrison was from a longstanding Strahan family and in the 1930s had joined a gang that logged the unique Huon pines that line the major rivers of the area. In 1940 he and his brother Ron achieved the near-impossible when they hauled a loggers' punt upstream on the Franklin through its largest gorge, the Great Ravine. After the Second World War, Morrison's connection with the area's wilderness spurred him to start the first regular tourist boat trips on the Gordon River. When the HEC's plans to dam the Franklin became known, he was horrified and started circulating anti-dam petitions on his boat, despite pressure from parts of the conservative tourism industry. During TWS's planning for the blockade he made the extraordinary offer that he would take all the blockaders from Strahan to a camp near the dam site on the Gordon River, some 50 kilometres away by water, without reimbursement. He told TWS: 'if you bring the people I'll run them up for free'.[43] His offer was crucial to the blockade; it was worth hundreds of thousands of dollars. Without it, the blockade would not have been the success it was, as Brown acknowledged: 'if Reg had decided to call it off we would have had no choice but to call it [the blockade] off'.[44]

The new year started well for Brown and TWS. On 1 January 1983, Brown was voted Australian of the Year by *The Australian* newspaper, then three days later he was released from gaol after reluctantly accepting bail conditions. The same day he was elected to state parliament, following the resignation of Sanders to stand for the Senate in an imminent federal election. It was extraordinarily symbolic for Brown to walk out of gaol into parliament and was the start of more than two decades of political life for him in state and federal arenas.

The second stage of the blockade started on 4 January, attracting an average of 50 new people each day in Strahan.[45] The tension around the blockade began to ratchet up. In early January new bail conditions were imposed on the blockaders that made it illegal for them to return to the west coast. Then, on 12 January, radio and telephone connections to the Strahan blockade centre were cut, and protesters were stopped by police from leaving the blockade campsite when the first bulldozer for the dam construction arrived in the town.[46] As passions increased, on 19 January Prime Minister Malcolm Fraser offered Gray the (then) enormous sum of $500 million if he stopped the dam, which Gray quickly refused.[47] Two days later the federal Labor parliamentary leader, Bill Hayden, visited the blockade and reaffirmed his party's commitment to stopping the dam if elected to office. The Franklin dam was locked in as a federal election issue with both national leaders knowing that popular opinion was firmly opposed to it.

The 1983 federal election and High Court action

In an attempt to catch the Labor Party off-guard, on 3 February 1983 Fraser called an early federal election for 5 March. Labor replaced Bill Hayden with Bob Hawke as its leader on the same day and quickly swung into the campaign. Brown told the national media that TWS (and the Australian Conservation Foundation which joined with it in a 'National South-West Coalition') would support the Labor Party, as well as the Democrats in the Senate. TWS and ACF also organised a large number of Franklin volunteers to campaign in seventeen marginal seats where Labor had a good chance of winning. This national single-issue marginal seat campaigning was a first in Australia.

On 4 February, 20 000 people attended TWS's 'Rally for Reason' in Hobart, making it one of the largest per capita rallies ever held in Australia.[48] On 8 February the third and final stage of the blockade began, culminating in a large action on 'G-day', the last day before the federal election media blackout began on 1 March (231 people were eventually arrested on that day). On 17 February the thousandth arrest at the blockade took place and, audaciously, a week later Gray made camping in the national park in which the blockade was situated illegal and many protesters were forcibly evicted.[49]

Although TWS felt it and the Labor and Democrat parties had run an effective election campaign, it was not sure Hawke would win the election. In an echo of the brazenly political role it played in the 1972 state election at the height of the Lake Pedder campaign, the HEC tried to help Fraser by placing advertisements in major newspapers defending the Franklin scheme. Not to be outdone, however, the National South-West Coalition placed the first-ever full-colour advertisements in *The Age* and the *Sydney Morning Herald* featuring photographer Peter Dombrovskis's unforgettable picture of Rock Island Bend on the lower Franklin. It showed a rocky island in the middle of the river surrounded by swirls of mist and water and was accompanied by the headline: 'Could You Vote For a Party That Will Destroy This?' Another image of the campaign that TWS pushed at every turn was its 'No Dams' triangular logo, the basic triangular form of which remains the logo of Green parties around Australia to this day.

Despite the uncertainty, in the end Labor was swept to power with a comfortable margin, winning 75 of the 125 seats in the House of Representatives thanks, in part, to the effective Franklin marginal seat campaign. As part of his election-night victory speech, Hawke reaffirmed that 'the dam will not be built'. Norm Sanders was also successful in being elected to the Senate. One disgruntled federal Liberal Party politician said of TWS's part in the Hawke victory: 'I cannot understand why an organisation with one thing on its mind should seek to turf out a democratically elected government'.

Hawke quickly made good on his Franklin promise and, with HEC earthworks continuing apace, at the end of March federal regulations forbidding dam works in the World Heritage Area were passed, backed by the passage of the federal *World Heritage Properties Conservation Act* in May (which was similar to earlier legislation introduced by the Democrats). The day after the regulations were put in place, the Tasmanian government made it clear it would not take the loss lying down, declaring it would challenge the validity of the legislation in the High Court, an action which the pro-development Queensland government joined.

The seven High Court judges considered the case from the end of May until mid June. The case largely turned on whether the federal government's signing of the international World Heritage Convention gave it the external affairs power to override the Tasmanian legislation. TWS had its own representation in the courtroom but was only allowed a short address to the court, and was not allowed to show the judges photos of the Franklin. The court's Chief Justice said they could 'inflame our minds with irrelevancies'.[50]

Brown and TWS grew frustrated with the legalities of the case, which seemed far removed from its perceptions of wilderness. TWS had no certainty of winning the case, although the court had upheld the federal government's use of external affairs power in its 1982 Queensland Koowarta decision, which allowed it to intervene in an Aboriginal issue in that state. The federal government also argued its power over corporations and race relations gave it the power to intervene. Hawke raised the stakes by quietly indicating to TWS that if the court action did not go the Commonwealth's way, his government would probably not pursue other measures to save the river. The night before the Court's decision was announced on 1 July, TWS knew it would either be ecstatic the next day or planning a continuation of the blockade. It would be heaven or hell. Brown talked about returning to gaol if the case was lost and secret caches of food were placed throughout the Franklin area to enable the blockade to be continued if need be. Heaven it was, though, when the Court announced it had decided by four to three to support the federal government's halting of the scheme. The minority included the Chief Justice, Sir Harry Gibbs. Many pro-dam sympathisers argued the narrow margin meant it was a dubious victory but the margin was large enough for Brown and TWS. The majority opinion was strengthened by the fact that three of the High Court's four judges had been appointed by past Liberal–Country party governments. However it was viewed, TWS had won an historic victory that more than vindicated the daring it had shown throughout the campaign. Brown called it 'a great day for Australia' and 'a peoples' victory'.[51] At a press conference following the Court's decision, he painted an expansive canvas of its significance:

> ... hope comes from the fact that the community made this change. [We have acknowledged] that there is a limit to the technological destruction we can allow in the name of progress ... I know that hope will go out from the decision to

> the people in the world who are fighting bigger issues, such as the nuclear arsenals that are aimed against the future of humanity, the imbalances between rich and poor . . . We have got some very big problems confronting us and let us not make any mistake about it, human history . . . is fraught with tragedy. It is only through people making a stand against that tragedy and being doggedly optimistic that we are going to win through. If you look at the plight of the human race it could very easily tip you into despair, so you have to be very strong. We all have to be because we are bearing up against a world political community and military that runs on ego. The people involved in this campaign by and large run on an ideal—the hope for a better human future—and it is not easy going all the time.[52]

Needless to say, there was bitterness amongst those who had wanted the dam. The head of the pro-dam group, The Organisation for Tasmanian Development, Kelvin McCoy, said: 'as far as I'm concerned, the rest of Australia doesn't exist'.[53] In an act of vandalism, a 3000-year-old Huon pine that was much visited by blockaders was destroyed. At the federal level, Hawke attempted to soften the blow to the state by immediately offering significant financial compensation to the Tasmanian government (which eventually came to $270 million).

Even now, the fight to save the Franklin River remains Australia's biggest environment campaign. Only a handful of campaigns—including those to save the Great Barrier Reef and Fraser Island in the 1960s, the Daintree rainforests in the late 1980s, and the ultimately successful campaign to stop the Jabiluka uranium mine in Kakadu National Park in the 1990s—came close to rivalling the Franklin battle. At the time, many of those people involved thought it would soon be eclipsed by bigger wilderness struggles but to this day it remains the biggest our country has seen. Its success was due to a combination of good management and, like so much of life, good luck. Campaigners learnt from many of the mistakes of Lake Pedder and the result was a well-managed campaign carried by one organisation, which was bold, tackled the economic case against the dam and went national early on in the fight. Lucky breaks included the knocking back of the Gordon-above-Olga compromise legislation by Tasmania's upper house; Lowe's decision to go ahead with the World Heritage nomination without the backing of his upper house; Fraser's refusal to pull the World Heritage nomination; Wran's preparedness to help swing a circumspect national Labor Party behind the Franklin cause; and the greater sophistication of print and electronic media at the time of the campaign.

A major enduring downside of the Franklin victory, however, was the further polarisation of Tasmanian society, a trend that started with the Lake Pedder controversy but got much worse during the Franklin campaign; the government of Robin Gray, in particular, never seemed to resile from using the issue to divide the state. However the victory is analysed, its significance remains with us today by forever putting mass environmental consciousness on the Australian map.

After the Franklin

The HEC was devastated by the Franklin result but wasted little time before regrouping. It quickly drew up a list of rivers outside the World Heritage Area that could still be dammed. To take advantage of the compensation money being offered by Hawke, one month after the High Court decision the Tasmanian parliament approved two more hydro schemes. One would dam the waters of the King River and another would dam the Henty and Anthony rivers. Both sets of dams were further from the heart of the state's south-west than the Franklin or Lake Pedder schemes were but both ate into the northern reaches of the wilderness much as the Pieman scheme had. By rights, like the Pieman scheme, the environment movement should have put more effort into stopping the King and Henty–Anthony schemes but campaigners were exhausted after the Franklin win and not at all confident the public would support them against more hydro schemes. Chris Harris, one of the directors of TWS who followed Brown, put some energy into investigating legal challenges against the new schemes but his efforts did not go anywhere.

The compensation paid by the federal government to the Tasmanian government mainly compensated it for the higher cost of the power generated by the King and Henty–Anthony schemes compared to the power that would have been generated by the Franklin scheme. And even though the HEC was keen to keep up a 'business as usual' attitude after the Franklin loss, enormous unforeseen changes soon beset it. The first was that the King and Henty–Anthony schemes ended up being the last schemes the commission ever built. Ultimately the high capital cost of its schemes, as well as the increasing scarcity of major unexploited rivers in Tasmania, made further hydro development unfeasible. Underscoring the financial vulnerability of the HEC's schemes was the fact that when the Pieman scheme was finished in 1987 it had a final cost of $691 million, more than four times the original estimate.[54] The last hydro power station in Tasmania opened in May 1994—just eleven years after the Franklin win. The HEC then scrapped all its dam construction staff and equipment. Its power was further diminished in 1998 when its distribution and retail arms were separated into different entities.

The Basslink cable

The threat posed to Tasmanian wilderness by the HEC (known as Hydro Tasmania since 2000) did not end with the opening of the King and Henty–Anthony schemes. In 1991 a national report by the federal Industry Commission recommended the creation of a national electricity market that would include a major power cable between Tasmania and the mainland.[55] In 1992 an environmental and social study into such a trans Bass Strait cable—'Basslink'—was released, followed by internal evaluation by the HEC four years later.[56]

On paper, the Basslink power cable made sense because it would allow Tasmania's hydro stations—which can be quickly turned on and off—to meet peak interstate power demand while the less flexible mainland coal-fired power stations could supply Tasmania's baseload demand. But apart from the cable's enormous cost, it posed a major threat to the Tasmanian wilderness with frequent water surges through the turbines of the power station that serves the Lake Pedder scheme. This would mean frequent fluctuations in the level of the Gordon River. These cold water flushings would scour the banks of the river, removing most of its sediment as well as associated plants and animals. The banks of the Gordon had already developed significant dead zones since the start of the Lake Pedder scheme but the Basslink scheme would make the sterile zones much more significant, roughly doubling their size. The flushings would denude much of the Gordon's riverine environment, something that was probably illegal under the 1983 *World Heritage Properties Conservation Act.*

The Greens, who since 1989 maintained a strong presence in Tasmanian politics, opposed the scheme. Their leader, Bob Brown, said: 'Basslink is a regressive idea that cuts right across Tasmania's clean, green image ... far from renewing nature, Basslink threatens Tasmania's wild rivers with twice-daily surges as the turbines are switched on to power Melbourne's toasters and air conditioners.'[57] Despite the opposition, however, the cable began operating in early 2006 as Tasmania's wilderness braced itself for yet another hydro onslaught. Basslink makes the possibility of one day draining Lake Pedder both more and less likely. It makes it more likely because the energy production loss of the draining could be covered by imported electricity. It also makes it less likely, however, because Hydro Tasmania's financial viability has become tied to Basslink and the need to import more power would hurt its bottom line.

Another twenty-first century challenge that confronted Tasmania's hydro system was declining rainfall, possibly caused by global warming. By early 2008 Hydro Tasmania's dams were only a quarter full raising the prospect of future power rationing.

By the late 1990s hydro industrialisation in Tasmania had come full circle: there was acceptance that no more major schemes would be built in the state and the HEC's workforce shrank from 5300 in 1985 to just 1709 by 1997. It went from being a dam construction authority to just an electricity generation agency. The HEC even began marketing itself as a provider of 'clean' electricity to the rest of the country. In 1998 it unveiled the first of several wind turbines it would go on to build. Meanwhile, the numbers employed by Tasmania's energy-intensive manufacturing industries kept falling. The number employed at the zinc factory in Hobart—the first operation to ever sign a bulk electricity contract in Tasmania—has fallen from 3000 in the early 1960s to a quarter of that number today. Many of the major electricity users in the state would not stay in business unless they continued to receive large subsidies via their cheap electricity tariffs.

The saddest legacy of Tasmanian hydro industrialisation is the long list of outstanding natural sites that it has obliterated, including the original Great Lake, the Shannon Rise, the Forth Falls Scenic Reserve, the original flow through the Cataract Gorge, Lake Pedder, the upper reaches of the Mackintosh and Murchison rivers and the upper reaches of the King, Henty and Anthony rivers. All were obliterated by the hydro juggernaut and all could have been saved had Tasmania pursued a more lateral and creative economic and energy path.

Throughout most of the twentieth century Tasmanians were generally in awe of hydro technology because of its apparent ability to create jobs, but by the start of the twenty-first century many felt

Cartoons by Ron Tandberg that appeared in *The Age* in 1982 and 1983 commenting on the Franklin campaign

jaded by it, realising it had created more problems than it had solved. Hydro power can be god-like, with the ability to divert whole rivers and flood vast areas of wilderness, but Tasmania has not used hydro technology carefully. It has left the state with a power-generating authority that barely manages to make money, the highest unemployment rate in the nation and a long list of destroyed wilderness areas. With great power inevitably comes great responsibility but Tasmania has rarely managed to use its hydro power with the humility it required.

FORESTRY

4

Sawmilling to industrial forestry

Tasmania is covered with a rich cloak of vegetation. Much of its landscape is spellbinding, speaking, as it does, of ancient botanical links to South America, New Zealand and Africa. It also speaks of the fact that until the start of the twentieth century much of the state escaped human development and essentially had not changed since the end of the last ice age. Tasmania's wilderness is a living museum. No part of it speaks of its primitiveness, and its innocence, as loudly as its forests. One cannot help feeling deeply moved when seeing a mighty Tasmanian *Eucalyptus regnans* tree standing the best part of 100 metres tall with an enormous girth, long peels of bark and towering crown clawing at the sky. If such evocativeness is powerful today, it was overwhelming to the island's earliest European visitors. When arriving at the Recherche Bay area in the state's south in 1792, French botanist and explorer J.J.H. de Labillardière exclaimed: 'The eye was astonished in contemplating the tremendous size of the trees, amongst which there were some myrtles more than 150 feet in height'.[1]

Where some see beauty others see opportunity, however, and Tasmania's forests have sparked some of its fiercest conservation battles. Although the fight for the island's forests had a low profile from the 1960s to early 1980s, for the last two decades it has been Tasmania's defining conservation issue. Forestry has a longer history in Tasmania than hydro development and is more tightly woven into its economic fabric, but Tasmania's forests evoke passions every bit as intense as those aroused over the state's wild rivers. Like the battle over hydro development, the battle over Tasmania's forests is inexorably linked to the past and an understanding of where the state's forestry industry has come from is essential to understanding today's forest clashes.

Early sawmilling

Like Labillardière, Tasmania's early European settlers were moved by its forests and it was not long before Peter Degraves established the state's first permanent sawmill on the slopes of Mount Wellington, behind Hobart, shortly after his arrival from England in 1824. Other sawmilling operations soon followed, particularly along the thickly forested banks of the Huon River in the south of the state. The enormous boost in Tasmanian economic activity that accompanied the 1850s Victorian goldrush added to the logging pressure on the island's forests. Further pressure came from the demand for timber exports to far-flung parts of the British empire, and from land clearance for agriculture which rapidly spread from the state's southern Derwent River and Midlands regions through to its thickly forested north coast.

Throughout most of the nineteenth century, there was little control over logging so early settlers cut down extensive stands of forest without caution. By 1885, 62 sawmills were operating throughout the island and it was not long before unsustainable overcutting was taking place.[2] Scientist Julian Tenson echoed the alarm felt by many when he warned that unless the state government took action 'the forests of Tasmania, peerless and priceless as they once were, will soon be things of the past'.[3] By 1881 it was clear the government had to act and the (ironically named) *Waste Lands Act* was amended so that licences could be issued for the felling of timber, royalties could be charged for the felling of trees and, most importantly, areas could be set aside for 'the preservation and growth of timber'.[4] Although a modest initiative, Tasmania's first legislative step towards preserving its unique forests had been taken. Four years later came a further significant advance with the passing of the first Tasmanian legislation dedicated to the management of forests that allowed for the appointment of a Conservator of Forests, who would take overall responsibility for state forestry. For the first time, someone was charged with taking a big-picture view of an industry that for too long had been conspicuously unregulated.

The inaugural Conservator of Forests was George Perrin, who ended up being the first whistleblower on the conduct of the island's forestry industry. By today's standards he could even be regarded as one of the state's first forest conservationists. Within days of his appointment, Perrin conducted inspections of the state's forests and was appalled by their wanton destruction. He saw extensive evidence of illegal logging and reported that the industry was in chaos.[5]

Perrin was a man ahead of his times. He recommended that a management regime be put in place that would include the setting aside of permanent forest reserves and the sustainable cutting of threatened species. But Perrin was too progressive and grew frustrated with the situation in Tasmania, eventually accepting an appointment as Conservator of Forests in Victoria.

In 1898, Perrin was recalled to Tasmania and this time he pulled no punches. Shortly after his return he boldly observed: 'I hold a particular concern for the insidious and cancerous destruction of trees . . . it is difficult now to realise the size of the trees that existed in the 1800s and the extent of the forests. It is desirable to step up a campaign to make the people aware of the dwindling and priceless heritage. Short-term thinking has beggared the forests of Tasmania.'[6] Stronger words could not be written today.

Despite Perrin's warnings, throughout the rest of the nineteenth century and into the early twentieth century the Tasmanian forestry industry remained chaotic and loosely regulated. Eventually, at the urging of a new Conservator of Forests, the state government established a dedicated forestry department in 1920 but that did little to fix the problem. In 1930 another forests conservator, S.W. Steane, repeated what Perrin had said four decades before, declaring: 'the regime of laissez-faire had become so firmly established that any mention of control or restriction was regarded as heresy of the most dangerous kind'.[7] Much of the laissez-faire attitude flowed from the granting of licences that gave logging companies exclusive access to extensive stands of forests, thus providing little incentive to manage them responsibly. These 'monopolistic leases' had originally been established in 1890; they eventually became known as forest 'concessions' and were the dominant forestry management regime for the next century. They gave forestry companies private landowner rights over public forests; a regime that worked against sustainable use of the resource. Dire though the situation was in Tasmania's forests, the pressures during the nineteenth and early twentieth century were modest compared to what was to come.

The arrival of the pulp and paper industry

Even though most of the forestry in Tasmania before the early twentieth century had been poorly managed, it had been relatively localised and did not penetrate far into the state's wilderness forests. It gnawed at the edges but went no further. What would significantly increase the impact of the state's forestry industry was the arrival of the pulp and paper industry. When the industry arrived the game changed: what had been haphazard, small-scale logging became reckless, large-scale industrial forestry.

Throughout the nineteenth century, paper had been generally made around the world from softwood timber pulpwood but, despite this, papermaking companies often inquired about the suitability of using Tasmania's hardwood forests. In 1914 the state government commissioned a United States scientist to look into the papermaking potential of Tasmania's forests but, like many before, he concluded they were not suitable. However, by the early 1920s promising experimental work had been done on the papermaking potential of hardwood forests and the Tasmanian government funded a new investigation of the suitability of its timber. The results indicated paper (particularly fine paper) could be commercially produced from the state's trees and the starter's gun was fired on a new age in Tasmania's forestry industry.[8]

Throughout most of the twentieth century, the Tasmanian government was keen to build its economic base by attracting manufacturing industries that depended on the use of natural resources. Tasmania's extensive stands of forests, vast stores of minerals and many wild rivers were reviewed as untapped wealth that could rescue the state from economic irrelevance. Despite a new global awareness of wilderness that began in the 1920s (partly as a result of the popularisation of the motor car), the Tasmanian government could only see unrealised economic potential in its wild lands. In 1916 it extended generous terms to enable the establishment of a major zinc works in Hobart (see Chapter 1) and a few years later saw a similar opportunity in the new hardwood papermaking industry. In 1924 state forestry minister Albert Ogilvie (who would later significantly expand hydro electric development in the state) said of the new pulp and paper industry: 'The establishment of this industry will have a far-reaching effect on the influence of this state and no better object-lesson could be pointed to as evidencing the potentialities of our timber'.[9]

To kick-start the new state papermaking industry in 1926, the euphemistically named *Wood-pulp and Paper Industry Encouragement Act* was passed. Two companies—Paper Makers Pty Ltd and Tasmanian Paper Pty Ltd—were given forest concessions across large areas of forest in the north-west and south-east of the state in anticipation of their building new pulpmills.[10] But the start of the Great Depression saw both fall on hard times and they amalgamated in

1936 to become Associated Pulp and Paper Mills Ltd (APPM). APPM went on to become one of the titans of the state forestry industry and a frequent target of conservationists' ire throughout the rest of the century. In 1937 it began operating a paper mill at Burnie. Industrial-scale forestry had arrived in the state.

The forests of north-west Tasmania were not the only areas eyed off for their papermaking potential. In 1922 the Tasmanian government investigated the forestry potential of the tall trees that grew in the Florentine Valley, west of Hobart, on the edge of the south-west wilderness. The Florentine Valley was home to the state's most extensive stands of towering *Eucalyptus regnans* trees and was the brightest of its tall forest jewels. *Eucalyptus regnans* can grow to over 100 metres and are the tallest flowering plants on earth; there was no richer store of them in Tasmania than in the Florentine.

Again, where some saw beauty others saw opportunity and the government was not content to leave them be, even though it had reserved the adjoining (modestly forested) Mount Field area as a national park in 1916. In 1932 the *Florentine Valley Wood-Pulp and Paper Industry Act* was passed, granting the Derwent Valley Paper Company access to the area's forest giants in return for an undertaking to establish another pulp and paper mill in the state. In 1938 the company became Australian Newsprint Mills (ANM), which included as major shareholders dominant Australian newspaper publishers Keith Murdoch (father of Rupert) and Warwick Fairfax (publisher of *The Sydney Morning Herald*). In 1941 the company began producing paper from its mill at Boyer near New Norfolk, west of Hobart, thus beginning the destruction of some of the island's most inspiring tall trees. Not content with the timber access already granted to the company, in the mid 1940s ANM began pressuring the state government for access to the forests that grew on the edge of Mount Field National Park, within the park's boundaries, next to the Florentine Valley. This brazen grab sparked the first major wilderness battle in Tasmania.

It is tempting to think of the conservation movement as only dating from the 1960s but, in fact, in Tasmania it had its beginnings two decades earlier, during the bruising battle over the Florentine Valley forests.

The fight for the Florentine and Mount Field forests

Although there are many national parks in Tasmania today, there were relatively few in the 1940s. None of the wild areas of western Tasmania were reserved in national parks or scenic reserves at the time, apart from the Cradle Mountain–Lake St Clair area in the centre of the state, the Frenchmans Cap area south of Lake St Clair and part of the Gordon River on the west coast. Given this, Tasmanians were proud of the fact that the Mount Field area had been the first national park created on the island, along with Freycinet National Park. News that ANM wanted to grab 1600 hectares of the park's forests came out informally, as is often the case in a small community such as that of Tasmania, reaching the ears of one of Tasmania's pioneering bushwalkers, Jessie Luckman, through contacts she had in the state bureaucracy and at a cocktail party she attended in 1943.[11] Luckman, and her husband Leo, were leading figures in the Hobart Walking Club and were both outraged that the state government could consider slicing off part of the national park to satisfy ANM's corporate hunger. Through her contacts Luckman learned exactly how much the government was secretly planning to excise from the national park and, after much persistence, got the Hobart *Mercury* newspaper to publish a letter from her calling on the government to come clean about its intentions for Mount Field.

When the government finally showed its hand, the first big public wave of conservation outrage in Tasmania broke. Harnessing that outrage was hard for Luckman and others who wanted to fight for the area's forests but did not know exactly how to go about it.

Conservation battles had not been waged before in Tasmania and they had few ideas about how to change the government's mind. Reflecting on the times, Luckman said: 'It was pretty hard going then because none of us were versed in campaigning and some did not even know how parliament worked'.[12] Undeterred, Luckman and fellow activists formed an action committee that became the Tasmanian Flora and Fauna Conservation Committee, the first-ever conservation organisation in the state and one that lived on until the late 1960s.

As was the case in other states, early conservation battles in Tasmania were mainly concerned with directly influencing politicians instead of trying to bring public pressure to bear on them as later battles would do. As part of their lobbying effort, the Flora and Fauna Conservation Committee took members of parliament on walks into the threatened forests and spent hours in parliament trying to learn its processes. The committee also sought information from overseas conservation groups such as the Sierra Club in the United States. But the trailblazing campaign was a challenge for all involved, especially when it came to interacting with the commercially driven government. Luckman said it was 'very hard to put your feelings into words that will count with people whose main aspect is the commercial side and the money'.[13]

In 1949 the legislation authorising the excision of part of the national park was forced through Tasmania's lower house of parliament but by the time it reached the upper house the Fauna and Flora Conservation Committee had sufficient confidence to be able to convince a parliamentary subcommittee that the bill should be thrown out. The bill's supporters, however, were ardent lobbyists and finally succeeded in getting it passed. Luckman and the committee were devastated. Despite all their idealism and hard work, the state's parliamentarians did not want to hear anything that might get in the way of the government's development vision. It was a bitter baptism of fire for the pioneering conservationists. Tellingly, the public works minister who chaired the parliamentary committee that endorsed the excision was Eric Reece who, as premier, would play a pivotal role in the revocation of Lake Pedder national park for hydro electric development two decades later.[14]

After the vote, the committee battled on and eventually forced the government to amend the excision so that the national park loss of the tall trees given to ANM was compensated for by lesser mixed forest added on the southern side of Mount Field.[15] The replacement forests did not compare with the ones given to ANM, however. Today precious few of Tasmania's *Eucalyptus regnans* forest giants remain; of about 100 000 hectares that existed when Europeans first arrived on the island only 13 000 hectares still stand.[16]

The most scandalous element of the Mount Field excision was the part played by corruption. ANM was mainly interested in the lower grade eucalypts while local sawmillers had their eyes on the area's large, tall trees. In the early 1940s the sawmillers bribed both the Forestry Department and the forests minister as part of their campaign to get their hands on the trees. A Royal Commission was held into the bribery allegations and its findings were published in May 1946. It found that sawmillers had offered, and given, bribes and that at least two Forestry Department officials, as well as the minister, Thomas D'Alton, had accepted the improper gifts. But none of them were found guilty of a criminal offence, partly because the Royal Commission found bribery was not a common practice of the sawmillers.[17]

No doubt the politicians Luckman tried so tirelessly to convince thought they were bequeathing the state a richer economic heritage than the natural heritage of the Florentine giants. But before the arrival of ANM, Tasmania had in the Florentine Valley a magnificent stand of trees that rivalled the giant redwoods of California. The politicians who gave them away had no idea what they were agreeing to. Today, if you climb one of the peaks in the western part of Mount Field National Park and look down on the

Florentine Valley, all you see is a maze of forestry roads, denuded slopes and feeble regrowth forest. It is very hard, these days, to conjure up what the Florentine must have been like before ANM arrived. The loss of the giant Florentine Valley forests rivals the loss of Lake Pedder.

The Mount Field National Park excision was a blow to conservationists but it was not the first time that part of a Tasmanian national park had been cut out to satisfy the forestry industry. The first occasion was in 1943 when 1214 hectares of Hartz Mountains National Park were excised after lobbying by logging interests. By the time of the Mount Field excision, no less than five national park excisions had already taken place: one in Hartz Mountains, two in Freycinet National Park in 1941 and 1942 (for mining) and two in Cradle Mountain–Lake St Clair National Park in 1939 and 1940 (for mining and hydro development).[18]

The battle over the Florentine forest giants showed how ill-prepared Tasmania was for the arrival of the pulp and paper industry. The state government was eager to embrace the industry but knew precious little about its long-term effects. Large tracts of forest were given over to the new pulp and paper monoliths without any certainty about the long-term ability of the forests to sustain the industry's level of logging. The Forestry Commission of Tasmania—the autonomous incarnation of the previous Forestry Department—was particularly ignorant about exactly how much cutting the state's forests could sustain. Although the concept of sustainable forestry yield had been around since the 1920s, before the Second World War the commission had little comprehensive data to base it on. It seems bizarre today, but the pulp and paper industry was given a blank cheque to fell forests without the government knowing how much forest resource there was.

After the war, however, a 'forest demarcation branch' of the Forestry Commission was established and it set about attempting to work out how much timber volume there was, basing calculations on aerial photographs taken for defence purposes during the war.

By 1951 the Forestry Commission was making its first estimates of the state's total commercial wood volume and, ominously, was also starting to make its first assessments of the maximum timber cut the forests could take. It estimated 30 per cent of Tasmania's commercial forests were accessible by road (or soon would be) and another 40 per cent could be reached with further road construction.[19] The first major plunges at the heart of the state's wilderness forests were being set up.

No sooner had the assessments of Tasmania's sustainable forestry yield been made than the first unambiguous signs emerged that some of the state's forests were, in fact, being overcut. After the Second World War there was a large increase in the amount of sawlog timber felled to satisfy a surge in construction activity, with the result that by the mid 1960s the industry was cutting two and a half times the volume it was cutting immediately after the war.[20] From the mid 1960s onwards, warnings by scientists and conservationists were made about the unsustainability of the sawlog cut but it was not until the late 1970s that they began to be taken seriously.

The pulp logs that fed the pulp and paper industry were viewed differently to higher grade sawlogs: there seemed no limit to the volume of this lesser wood that could be taken. As a result, in 1959 an act was passed for the pulp and paper industry to set up in the Huon Valley, in the state's south-east, and a pulp pelletising plant was established at Port Huon, commencing operation in 1962. The plant drew on pulpwood logged in the eastern fringes of the south-west wilderness. In 1961 yet another pulpmill was authorised for APPM, this one situated at Wesley Vale on the mid north coast of the state (although only a small scaled mill ended up being built there). Within 25 years of the establishment of the first pulp and paper mill at Burnie, no fewer than four such mills had set up in the state. Increasingly, it was no longer timber-felling for sawlogs that drove the forestry industry in Tasmania but the huge pulpwood hunger of the new pulp and paper mills.

The arrival of the woodchip industry

The new, corporate hunger for Tasmanian pulpwood did not stop with the Wesley Vale mill. From the early 1960s Japanese paper executives began visiting Tasmania, expressing interest in using the state's forests as a source of export woodchips to feed their paper mills in Asia. They were not keen to set up downstream processing mills in the state that might convert timber to pulp or paper, as the pulp and paper industry had, they just wanted to ship out crude woodchips. As with the pulp and paper industry, the Tasmanian government welcomed their approach, again paying little heed to the possible long-term effect of the new industry on the forest resource. The Forestry Commission shared in the excitement of setting up an export woodchip industry and in 1968 reported that the new industry would transform 'stagnating' forests, making them into 'continuously productive stands not only of pulpwood but also of sawlogs'.[21] Despite the 30-year experience of the pulp and paper industry, no caution accompanied the setting up of the woodchip industry; it was as if the state had learnt nothing about sustainable forestry. If anything, the Tasmanian government was more compliant with the new woodchip operators than it had been with the earlier pulp and paper companies. When forest concessions over vast areas of state forest were granted to the pulp and paper companies they were expected to establish a mill that would add value to the resource. But the woodchipping companies were expected to do virtually nothing in return. The Tasmanian government was happy for them to reduce the state's forests to minute pieces then ship them overseas without any value-adding whatsoever. The most it expected was for some of the woodchip operators to look into the feasibility of establishing a local mill but if the company thought such a mill was unviable it could still have the woodchip timber.

Many justifications were concocted for the new woodchip industry; two of the more popular ones were that it would make for more efficient use of the forests—since nearly all the timber would now be removed instead of just the sawlog resource—and that forest regeneration could be more effectively carried out on clear-felled land than in selectively logged forests. Whatever the justifications, from the late 1960s onwards a steady procession of new woodchip operators gained a foothold in the state's forests. In 1968 a new company, Tasmanian Pulp and Forest Holdings (TP&FH), was given permission to set up a woodchip mill at Triabunna on the state's east coast. Soon after, APPM was given permission to set up a woodchip mill, in addition to its pulpmills, which it did in 1972 on the Tamar River in the state's north-east. This followed the establishment of Australia's first woodchip mill, at Eden on the south coast of New South Wales, two years earlier. Another woodchip mill was also given permission to set up on the Tamar. This one would be operated by Northern Woodchips, which became Forest Resources. Unlike the other two woodchip mills, this operator would source all its pulpwood from private land and sawlog waste.[22]

Inevitably, the arrival of the woodchippers hugely increased the pressure on Tasmania's forests and therefore the threat to its wilderness areas. Between the late 1960s and the mid 1970s the amount of sawlog removed each year from state forests remained largely unchanged while the amount of lower grade pulpwood cut for the pulp and paper industry increased by about a third (to 600 000 tonnes). But the amount of pulpwood cut for the new woodchip industry grew exponentially, from nothing to 1 000 000 tonnes over the same period (and that was without counting pulpwood cut on private land).[23] The days of small-scale, localised forestry were well and truly over in Tasmania. Big-business forestry, driven by global markets, had arrived and the forests would never be the same again. The threat posed to the state's wilderness was also on a whole new scale. Woodchipping would go on to be public enemy number one to the state's conservationists.

The big change that woodchipping brought to Tasmanian forestry was a large increase in clearfelling: the flattening of all trees in an area logged for their timber. Clearfelling was not new—it had started with the advent of the pulp and paper industry in the state in the late 1930s—but the new woodchip industry made it all-pervasive. It became the ugly face of Tasmanian forestry. The more entrenched woodchipping became, the more people saw denuded slopes from their cars and from lookouts. By the 1980s more than 20 000 hectares of forest was being clearfelled in the state each year. More than anything else, clearfelling came to symbolise all that was wrong with woodchipping and Tasmanian forestry in general. The scorched-earth vistas left behind by clearfelling were a public relations disaster for the industry. In the late 1980s a Tasmanian forestry minister attempted to appease the public by telling them: 'clearfelling is ugly to the untrained eye'. One pro-forestry Labor member of federal parliament even tried to justify it in 2007 by saying: 'forestry has the same problem as abattoirs—we all like the end product but it doesn't look very good'.[24] These proponents were fighting a losing battle; clearfelling would always look ugly.

An early book that rang warning bells about clearfelling was *The Fight for the Forests* by Ron and Val Routley, published in 1975. The Routleys warned of the long-term consequences of clearfelling, noting that: 'very little is known of the regenerative properties under clearcutting of most of the eucalypts',[25] and declaring that 'the forested part of Tasmania is regarded as one of the last great wildernesses of the world. Soon it will cease to be.'[26] The Routleys speculated that the long-term impacts of clearfelling could include high soil erosion, loss of soil nutrients, destruction of forest animals (through habitat destruction and greater competition for remaining forest habitats) as well as increased risk of pest and disease introduction. Their words were prescient.

Soon it became obvious that woodchipping would place unsustainable pressure on the state's forests. As early as 1976 the Tasmanian government became concerned about the impact the industry was having on forests on private land, establishing an inquiry into private forestry which recommended greater oversight.[27]

It also became obvious that a vast area of forest would be needed to support the combined onslaught of the woodchipping, pulp and paper and sawlog industries. By the late 1970s the Forestry Commission was setting a target of 4 million acres (1 618 000 hectares) of public forest—equal to just under one-quarter of the state—to satisfy the combined timber hunger of the industry.[28] On top of this, a further 1 million hectares of forest on private land would be utilised for the same purpose. When combined with the forests on public land, this would equal just over one-third of the state. Much of that third went right up to the edge of Tasmania's wilderness areas. Forestry bit at the largest contiguous stand of temperate rainforest in Australia in the state's north-west, 'The Tarkine'; it went right up to the windswept ramparts of the alpine Central Plateau in the middle of the state; it despoiled the northern edge of Macquarie Harbour on the west coast; and it ate into the long eastern boundary of the state's south-west wilderness. From the 1970s it was everywhere.

The first anti-woodchipping campaign

The 1970s and the first half of the 1980s were dominated by the fights to save Lake Pedder and the Franklin River so the campaign to save the island's forests largely took a back seat. That did not stop some early protest action against the new woodchipping industry, however. One of the areas where woodchipping first made its presence felt was along the Great Western Tiers, on the northern edge of the Central Plateau. In a small town called Meander, a year-long campaign was mounted by a group of activists against the woodchipping operations of APPM, starting in 1974. They were appalled by its large-scaled clearfelling that flattened much of the old-growth forest of the area. They published a pamphlet and a poster that declared: 'Woodchip laughs at you all the way to the bank!'[29] They lobbied the union movement and went on talkback radio. They also gave evidence to a 1975 Senate inquiry into wood-chipping. Like the activists working with Jessie Luckman in the 1940s, the Meander activists were new to campaigning and often felt unsure about what they were doing. It was a small effort compared to the bigger campaigns that would follow but it showed APPM that people were prepared to fight for the forests.

The Hartz Mountains forest swap

The 1940s fight by Luckman and the Tasmanian Flora and Fauna Conservation Committee for the tall trees in Mount Field National Park had been a bitter and disappointing affair for the Tasmanian conservation movement, but it was not an isolated event, as already noted. And it was echoed three decades later in another controversial carving-out of national park forests. This time it was another excision of forests in the Hartz Mountains National Park in the state's south-east. The story of the 1970s excision of the Hartz Mountains forests began in 1972 when a number of Tasmanian conservation groups—including the Tasmanian Conservation Trust and the South West Committee—lodged an objection to the granting of a mining licence to a company called Mineral Holdings to mine for limestone at Precipitous Bluff. Precipitous Bluff is in the south-west, a towering, rocky mountain near the area's south coast that was then outside the Southwest National Park. The mining company lost the hearing but won appeals to the Supreme Court of Tasmania in 1973 and 1975. The company also went on to win an appeal to the High Court in 1977 (see Chapter 8). During all the legal wrangling, the fight for Precipitous Bluff assumed a high profile—bumper stickers declaring 'Save Precipitous Bluff' and 'Don't Scruff our Bluff' adorned Tasmanian vehicles while many newspaper articles were written on the issue. Coming a few years after the battle to save Lake Pedder and a few years before the fight for the Franklin River, the fight for the bluff took place during a time of heightened environmental awareness when there were no competing environmental issues. At the same time, the state government was considering expanding the Southwest National Park.

Whether Mineral Holdings could ever have profitably mined at Precipitous Bluff was doubtful; the area was a long way from the nearest road and had no nearby shipping infrastructure. But whatever the company's motivation, the Precipitous Bluff issue caught peoples' imaginations and became a cause célèbre. In the midst of all the agitation, Australian Paper Manufacturers (APM) saw a golden opportunity and put a case to the state government that the forests around Precipitous Bluff—that lay within the large forest concession the company controlled—could be preserved in an enlarged Southwest National Park in return for a 'swap' of 2150 hectares of prime forest in the western part of Hartz Mountains National Park. It was a copy of the 1949 excision of Mount Field's forests that ANM had successfully engineered, and the Tasmanian government once again complied. This was despite the lack of any obligation to offer any compensatory forests to APM under the terms of its concession. To make matters worse, some people in the conservation movement—the Tasmanian Conservation Trust in particular—gave tacit support to the move, causing a major schism within the movement. The activists who supported the move were small in number, and were exhausted by the unsuccessful legal battles over Precipitous Bluff, but to this day their support is controversial and is sometimes thrown up as an example of what can happen if the environment movement attempts to compromise too much.

In October 1976 the legislation excising the forests was passed. Many activists, including long-time campaigner Helen Gee, argue the swap was a 'million dollar windfall' for APM[30] because, like Mineral Holdings, its resource around Precipitous Bluff was isolated and probably unprofitable to access. Yet again, national park protection had proven to be a toothless tiger. It had not saved

Hartz Mountain's or Mount Field's forests in the 1940s, it had not saved Lake Pedder in the 1960s and now it was failing to save the Hartz Mountains forests in the 1970s. Tasmania was developing an infamous reputation for opportunistic destruction of invaluable wilderness areas previously thought to be safe within national parks.

The Hartz Mountains swap was the first big issue fought by the Tasmanian Wilderness Society (TWS) that was formed in 1976 out of the South West Action Committee and which went on to successfully fight for the Franklin River (see Chapter 3). The year after TWS was formed, the first logging bridge was built across the Picton River near the excised Hartz Mountains forests. The Picton River had long been a barrier between the logging that went up to its eastern bank and the pristine forests on its western bank, but in 1977 the forestry industry wanted to breach it and strike further into the wilderness on its far shore. TWS placed an advertisement against the bridge in the local newspaper, the *Tasmanian Mail,* and lobbied the state government.[31] Bob Brown and Kevin Kiernan from TWS went to see forests minister Neil Batt (who had been instrumental in the Hartz Mountains forests swap) but got nowhere with him. TWS considered employing direct protest action against the bridge—of the sort later successfully used on the Franklin and in forest campaigns—but action like that was rare in those days and the idea was dropped. Brown later said: 'we thought about direct protest action that had been considered at Pedder but there were no models really and the organisation was not there'.[32]

The fight for the Denison River Huon pines

Despite the arrival of the forest-hungry woodchip industry, and the Hartz Mountains swap, the 1970s were not all about defeats in the fight to save Tasmania's forests. A major victory was chalked up in 1970 when, thanks to the dogged persistence of one man,

400 hectares of Huon Pine forest was preserved on the remote Denison River in the heart of the south-west. That man was Olegas Truchanas, the famous wilderness photographer and campaigner who was also mentor to Peter Dombrovskis, another photographer who would later become famous for his Tasmanian wilderness photography. Truchanas did a lot of canoeing trips into remote parts of the western wilderness and was aware of an extensive stand of unique Huon pines—much favoured for boat-building, furniture-making and arts and crafts—on the Denison River. In 1968 and 1969 he lobbied the Australian Conservation Foundation and the Tasmanian Conservation Trust about creating a reserve for the Huon pines.[33] As a result of his approach to the trust, he was given the chance to lobby the minister for forests but the Forestry Commission argued the pines should not be preserved because there were alternative stands that could be reserved. Truchanas was not easily put off, however, and miraculously was able to arrange an inspection of the commission's claimed alternative forests with a helicopter supplied by the Hydro Electric Commission. The inspection showed there were no major forests of Huon pine where the Forestry Commission claimed there were and after further inspections, and a lot more persistence on Truchanas's part, the Denison River Huon Pine Scenic Reserve was finally gazetted in August 1970. In contrast to the mass agitation that accompanied the fight for Precipitous Bluff, Truchanas won the preservation of the Denison River Huon pines by himself and the victory was a testament to his legendary doggedness.

Another local forest campaign carried on throughout the 1960s and 1970s was a fight to preserve the Norfolk Range forests in the far north-west of the state. One of the principal activists in that fight was Peter Sims, who played a significant part in the Lake Pedder campaign. Sims and others made several trips into the remote Norfolk Range area from the mid 1960s onwards and, like Truchanas, lobbied the Tasmanian Conservation Trust to become involved.[34] For Sims, the Norfolk Range area was one of 'the key wilderness areas left in Australia'.[35] Sims and fellow activists got local sawmillers to publicly agree that there was no commercial timber in the area and even got sheep graziers, who ran stock on the coast near the range, to agree to relinquish their leases. An environmental study was undertaken and the Norfolk Range came tantalisingly close to being gazetted as a national park before the state government changed in 1972, returning the Labor Party under Eric Reece (who had been responsible for the flooding of Lake Pedder). The Reece government dropped the national park idea, instead opting for lesser 'Protected Area' conservation of the area, much to the bitter disappointment of Sims. Unbeknownst to Sims at the time, this was only the beginning of the battle for the north-west forests which continues to this day.

Although inspiring, in general terms the forest battles waged by Truchanas, Sims, TWS, the Tasmanian Conservation Trust and the Meander anti-woodchipping activists were the exception. The two decades from the early 1960s to the early 1980s were dominated by the fights to save Lake Pedder and the Franklin River and not by forest battles. This was not a deliberate strategy on the part of the environment movement; rather, it had much to do with the priorities of the media and the immediacy of the dams planned for the Gordon River. Forest campaigns stayed in the background, although it is wrong to think that the Pedder and Franklin campaigns had nothing to do with forest preservation. Both campaigns put Tasmania's wilderness on the map and heightened public awareness of the uniqueness of the state's forests. The Franklin campaign also established the World Heritage Area preservation mechanism that would later prove invaluable to the Tasmanian forest campaign. A major figure in the fight for Tasmania's forests from the mid 1980s onwards, Geoff Law, argues:

> The fact is that by saving the Franklin we gained the World Heritage mechanism, national public awareness of the south

> west, and a powerful profile for the environment movement which eventually translated into success on the forests front. Without the success on the Franklin it would have been just a demoralising piecemeal grind, area by area, a coupe here and a valley there, and all these separate names that would have meant nothing to the general public.[36]

Until the mid 1980s, forestry in Tasmania was a sleeping giant left largely undisturbed by the conservation movement. As a result, destructive forest industry practices too often went unchallenged. Soon after the Franklin campaign, however, the forestry giant was stirred and would never sleep again.

5

Farmhouse Creek to the Salamanca Agreement

By the mid 1980s the innocence of both the Tasmanian environment movement and Tasmania's wilderness forests was well and truly lost. By then the conservation movement had the bruising battles over Lake Pedder and the Franklin River under its belt. It had also been half a century since industrial-scale forestry had arrived in the state with the establishment of its first pulpmill in 1937. People still stood in awe in Tasmania's forests, but it was with increasing apprehension that the forests might have a limited future.

After saving the Franklin River in 1983, campaigners in the Tasmanian conservation movement, particularly the Tasmanian Wilderness Society (TWS), were exhausted. For several years the Franklin campaign had been carried on at a punishing pace and the movement's campaigners no longer had much fight left in them. Most had worked as volunteers during the Franklin campaign and had put their lives on hold while it was fought. TWS was a young organisation and had been stretched to the limit by the Franklin.

Soon after the Franklin celebrations many in the movement came to realise, however, that the fight for Tasmania's wilderness areas had only just begun and that the Franklin campaign had been an opening act instead of a grand finale. The threat posed by the state's forestry industry had lurked in the background during the Franklin campaign but once it was over a feeling took hold that forestry was the next great front. Bob Brown summed up the mood:

> I came out of the Franklin campaign in 1983 with forests blinkering all around the place and I thought, 'I can't face this.' The dam was a single entity; a single set of job figures, a single set of economic figures, one suite of environmental values fitting into the heartland of wilderness. Forestry was all over the place, it was most complicated. We had no really good information on the economics of the industry . . . I thought, 'this is just too much . . .!' But there was no getting around it. Everywhere you went the wilderness forests were under threat around the state. So into the forests we went.[1]

The woodchip environmental impact statement campaign

The issue that pushed the Tasmanian conservation movement into the forests was the renewal of the state's woodchip licences, flagged by the federal government in late 1983. The movement was unprepared to campaign against the woodchip licence renewals but knew it had to somehow tackle the forestry industry over them. The Tasmanian export woodchip industry at the time was controlled by three companies: Associated Pulp and Paper Mills (APPM) and Tasmanian Pulp and Forest Holdings—both eventually owned by North Broken Hill—as well as Forest Resources, owned by Petersville-Sleigh.[2] At first it seemed the federal government would simply rubber-stamp the new woodchip licences but after a hastily organised 'write in' campaign, the government of Bob Hawke agreed to conduct an environmental impact statement (EIS) process in relation to the licences. An EIS would assess the full environmental impact renewal of the licences would have on Tasmania's forests.

The Hawke government also agreed to place a temporary moratorium on logging in two iconic areas that faced imminent threat from woodchipping, the Lemonthyme forests near Cradle Mountain, in the state's north, and the Farmhouse Creek forests in the Picton Valley on the eastern edge of the state's south-west wilderness. There was now a specific threat that had to be fought and a specific time frame within which to fight it. The movement had no alternative but to dive into the forestry issue in a big way.

The two big challenges for the state conservation movement in 1983 and 1984 were to raise community awareness about forestry and to prepare for the bureaucratic EIS process. In many ways the wordy paper war that the EIS process became was the antithesis of what the Tasmanian environment movement had been about during the Franklin blockade. Instead of major confrontations between police, workers and machinery on the one hand and idealistic greenies on the other, the movement was thrown into a bureaucratic process carried on well away from the glare of news cameras and reporters, where the central players were unseen forestry company executives and dour government bureaucrats. Although cumbersome and often uninspiring, the EIS process was, at least, a useful means through which the environment movement could learn about the forestry industry and raise public awareness about the threats it posed.

A second major post-Franklin TWS wilderness campaign at the time was its pitch for a 'Western Tasmania National Park'. This was a blueprint for significantly expanded boundaries for the national parks that covered the south-west. Launched in March 1984, the proposed boundaries extended across to the west coast, down to the middle reaches of the Huon, Picton and Weld rivers and north to the Great Western Tiers. It was a bold statement by the environment movement that did much to define the scope of the forests it campaigned for.

Another significant post-Franklin conservation movement forest campaign initiative was the creation of the Forest Action

Network, a coalition of the state's forest conservation groups. The network had a triumvirate of aims: greater forest reservation, improved forestry industry management and better industry development resulting in more employment. Like the Western Tasmanian National Park proposal, the network's aims helped frame the conservation movement's forest campaign as something broader than just a response to the woodchip EIS.

After being delayed for months, the draft woodchip EIS was unveiled in Hobart and Canberra in March 1985. It was soon obvious the document pandered to the forestry industry and had scant regard for the protection of the state's wilderness forests. It even claimed: 'the woodchip industry can be maintained at its present level without adverse effects on the environment'.[3] In response, conservation campaigners held the first major street rally in Hobart since the Franklin campaign[4] and immersed themselves in the process of writing an authoritative response to the draft EIS.

Although the movement managed to do a lot in a short period, its EIS campaign never came close to matching the passion the Franklin campaign had aroused. As a result, politicians in Canberra—in whose hands the fate of Tasmania's forests ultimately lay—did not feel pressured by it. Few were surprised then when, at the end of 1985, the Hawke government handed the Tasmanian forestry industry virtually everything it wanted. The federal primary industries minister, John Kerin, maintained that the federal government could not intervene in Tasmania's forests and successfully urged cabinet to extend the woodchip licences for fifteen years.[5] Federal environment minister, Barry Cohen, recommended that significant forest protection should be part of the woodchip decision but his advice was largely ignored by Kerin. In response, Bob Brown promised the movement would 'raise a campaign to save Australia's forests from Daintree [in Queensland] to Tasmania'. The day after the decision a huge banner was erected on a Hobart chimney tower that read 'Woodchipper Hawke'.[6] A large pro-woodchip rally was held by loggers in Hobart and in response TWS hired an aeroplane to fly over the city trailing a conservation message behind it. TWS went on to describe the EIS as the 'most transparent piece of industry propaganda ever prepared'.[7]

The crucial forests that the conservation movement wanted protected from woodchipping were those listed on the Register of the National Estate, managed by the Australian Heritage Commission in Canberra. The register was originally established by the Whitlam government as a national inventory of unique natural or cultural sites. If a site on the register was threatened by development, the federal government was obliged to explore all 'feasible and prudent alternatives' to the proposed development before allowing it to go ahead.

In the 1990s the focus of the Tasmanian forests campaign changed, slightly, when the movement began campaigning for the remaining old-growth forests of Tasmania to be protected but throughout the 1980s the state's National Estate forests were its main game. Although it was a relatively bureaucratic way of defining the state's high conservation forests—and not a definition that most of the public could readily identify with—the National Estate label was nonetheless a neat way of identifying the forests that had to be saved. These forests only represented about 10 per cent of the island's commercial low-grade timber resource so the movement always felt they represented a reasonable claim. The specific National Estate forests of most importance were the forests of the Great Western Tiers, which formed the northern slopes of the Central Plateau; the Lemonthyme forest near Cradle Mountain; the forests around the Douglas and Apsley rivers on the central east coast; and the forests that ran along the eastern border of the state's south-west. These areas were the most significant in terms of area and timber resource and were therefore the most contentious.

The all-important clause missing from the Hawke government's 1985 woodchip licence decision was a firm undertaking to protect the National Estate forests. When the federal government finally signed off on the decision there was a vague commitment to protect 'values' of the forests but nothing solid.[8] The conservation movement knew this, as did the Tasmanian forestry industry. To test the decision the forestry industry, with the backing of the Forestry Commission of Tasmania, sent bulldozers and chainsaws into the Lemonthyme and Farmhouse Creek forests in February 1986—straight after a state election. But they miscalculated. All hell broke loose.

The Lemonthyme and Farmhouse Creek blockades

Once it became clear the loggers were going to move into the Lemonthyme near Cradle Mountain and Farmhouse Creek forest in the Picton Valley, the conservation movement began planning for blockades at both sites. This was a big step: the blockades would be the first major forest direct actions held in Tasmania and were the first significant environmental direct actions in the state since the Franklin blockade. The loggers moved into both areas in February 1986 and were immediately confronted by the blockaders. Once again, people were pitted against machinery, as they had been during the Franklin blockade. On 25 February, 30 protesters stopped the progress of a bulldozer in the Lemonthyme. Meanwhile, at Farmhouse Creek, Alec Marr, subsequently a major figure in TWS, set up camp 30 metres up a tree in the middle of the loggers' path.

As more and more people took part in the blockades tensions grew between the protesters and the loggers. It was not long before it turned ugly—very ugly. Within a fortnight a major confrontation took place at Farmhouse Creek when two busloads of loggers came down to the blockade site to confront the 100-odd protesters camped at the creek. Fortunately, the blockaders had received an anonymous tip-off the night before so TWS ensured as many people as possible were in the camp and that the media were present to witness any scuffles. The police also sent along about 40 representatives.[9]

Shafts of morning sun burst through the sentinel trees of Farmhouse Creek on 7 March 1986 but the dawn chatter of birds was soon drowned by the din of the massing of loggers, police and machinery on the northern side of the creek. Meanwhile, blockaders braced themselves on the southern side. The loggers' advance was spearheaded by two large D9 bulldozers that headed for the creek and the blockaders. As the bulldozers crushed their way forward many of the blockaders moved into the stony bed of the creek that was an important border for them between wilderness and development. When the bulldozers reached the creek, a fast-moving melee broke out. A number of protesters jumped onto the bulldozers while others, including Bob Brown, curled up underneath so the dozers could only advance by injuring them. Burly loggers moved in and started threatening the protesters with chainsaws. One even began cutting the tree Marr was in. Finally, the loggers decided they would forcibly clear the way and began rough-handling the protesters. They grabbed, pulled, punched and yanked the blockaders, many of whom screamed out in pain.[10]

The loggers' prize was Brown. Four loggers grabbed each of his arms and legs while his shirt was ripped apart. They showed no mercy and were determined to mete out rough justice to him. Unfortunately for the loggers, the media captured the incident on film and the rough-handling of Brown became an iconic image of the blockade and of Tasmanian forest campaigns in general. He looked Christ-like pitted against the might of the helmeted loggers and the giant bulldozer. Imagery is all-important in modern-day conservation battles and the loggers' treatment of Brown unwittingly handed the conservation movement a major visual prize. All the while the police stood and did nothing; one later told Brown they had been ordered to take no action if a major conflict took place.[11]

Soon after, the state government, led by Robin Gray—the same premier who had championed the Franklin dam—invoked new 'exclusion' laws to keep the media and protesters out of the forests. Desperate for justice, several months later Judy Richter, from TWS, took a civil action against the manager of the company that sent the loggers in. The manager was convicted of assault, although he was only given a suspended sentence.[12]

Two days after the 7 March confrontation, shots were fired in the vicinity of Brown, Richter and a journalist at Farmhouse Creek. Brown summed up the campaign's significance: 'The battle of Farmhouse Creek will ring down through the generations as a watershed in the campaign to save the forests, not only of Tasmania, but throughout Australia'.[13]

In the end, the police did act at both Farmhouse Creek and Lemonthyme but not to protect the protesters. In mid March they moved into both camps and arrested 70 people, effectively ending both blockades.

Although the conservation activists were forced out, the blockades were a major win for the environment movement. Its cause was in the media, again, in a way it had failed to be since the Franklin days. The confrontation stimulated major media coverage of the issue and a 3000-strong conservation rally was held outside Tasmania's Parliament House. Farmhouse Creek and the Lemonthyme became household names and, finally, the movement had a well-known conservation issue to pressure politicians with. The ambushing of the Farmhouse Creek blockaders was a tactical blunder that the loggers would never repeat.

An influential federal politician who was eager to listen to the movement's newly emboldened forest message was Graham Richardson. A 'fixer' in the Hawke government, Richardson was keen to build bridges with the environment movement, partly because he figured its appeal to marginal voters could have a pivotal influence on federal election outcomes (which was the case in the 1990 federal election). Richardson had recently become federal environment minister and in many ways his affiliation to the right wing of the Labor Party made him an unlikely ally.

Richardson had been to Tasmania to inspect the forestry operations of pulpmill operator Australian Newsprint Mills, so the month after the Farmhouse Creek blockade drama he accepted an invitation from TWS to hear its side of the argument. He ended up flying over, and landing in, the forests of the south-west with Brown, Geoff Law (then with the Australian Conservation Foundation) and David Heatley (from TWS). Richardson was overwhelmed, declaring that by the time he got back to Hobart he was a 'convert' to the conservation cause.[14]

Richardson's influence soon became apparent. When the final woodchip licence agreement was signed between the federal and state governments in June 1986, Hawke emphasised his commitment to protecting National Estate forests. At the end of the year he announced a public inquiry into the protection status of Tasmania's National Estate forests.[15] TWS highlighted the importance of the Lemonthyme National Estate forests by holding a rally at Cradle Mountain in December 1986. Hawke's public inquiry followed a federal interdepartmental review that concluded logging should stop in the Lemonthyme but could proceed in the Jackeys Marsh–Quamby Bluff area of the Great Western Tiers in the state's north.

The catalyst for Hawke's review was a brazen directive from Gray that logging should commence in the National Estate forests of the Jackey's Marsh–Quamby Bluff area (although some argued Hawke was using the inquiry as a way of deferring the issue until after the 1987 federal election). Gray's edict was issued only six months after Hawke's National Estate forests undertaking. Hawke called on Gray to stop the logging while his government's twelve-month review was carried out, but Gray refused. Hawke had federal legislation passed to give effect to his Tasmanian forests inquiry, as well as a moratorium on the logging of the National Estate forests, but Gray still refused to cooperate. He only bowed to Hawke's wishes after the federal government won a High Court case in May 1987.

Gray was as unstinting in his support for the forestry industry as he had been for the construction of the Franklin dam. In 1985 he echoed his earlier claim that the Franklin River was nothing more than a 'brown, leech-ridden ditch' by declaring that 'prior to the advent of the woodchip industry the east coast forest areas were covered largely with decadent and rotting forests'.[16]

The Helsham Inquiry

The Hawke government decided to run its Lemonthyme and Southern Forests Commission of Inquiry much like a court hearing, choosing a retired New South Wales Supreme Court judge, Mr Justice Helsham, to head it. The inquiry focused on whether the National Estate forests of the Lemonthyme and Southern Forests (a generic name for the forests that bordered the south-east corner of the Franklin-era World Heritage Area) were worthy of World Heritage listing. Assisting Helsham was Robert Wallace, an economist, and Peter Hitchcock, a former forester with an environmental science background. The federal government gave financial support to the conservation movement to help it prepare its inquiry case. With the money, the movement was able to set up a dedicated office employing about half a dozen people under the direction of Alistair Graham, who went on to become Director of TWS.

The Helsham Inquiry represented a return to a fairly dry, bureaucratic form of campaigning for the movement, conducted mostly away from the gaze of the media, much as the 1985 woodchip licence renewal EIS process had been. The inquiry's first hearings were held in June 1987. Predictably, the Forestry Commission of Tasmania wanted to continue logging in the forests in question while the inquiry was being held and even had its case heard, unsuccessfully, by the High Court in May 1987 (the same case that forced Gray to cooperate with the Helsham Inquiry).[17]

For the conservation movement the Helsham Inquiry represented a temporary narrowing of its forests campaign. Its focus shifted from National Estate forests across the state to the World Heritage value of the specific wilderness forests of the Lemonthyme and Southern Forests areas. This was done so that the federal government might use its World Heritage powers to intervene in the forests debate.

At the start, the conservation movement had faith in the ability of the Helsham Inquiry to provide an umpire's view of how much forest should be saved. As the inquiry progressed, however, the movement lost faith and by early 1988 was circumspect. Alistair Graham came to feel that, despite its pretence of intellectual rigour, the inquiry was basically 'an enormous political event'.[18] In May 1988, when the inquiry's findings were released, there was little surprise when Helsham and Wallace said only about 10 per cent of the National Estate forests were worthy of World Heritage listing. Hitchcock disagreed. He concluded that all the forests the inquiry considered, and more, were worthy. Nine of the eleven expert consultants hired by the inquiry also distanced themselves from Helsham's and Wallace's findings.[19] Brown called the inquiry's finds 'a win for the companies, the chainsaws, the cable loggers, the woodchip mills'.[20] TWS and the Australian Conservation Foundation (ACF) mounted a major campaign in response to the inquiry's findings that included a large rally at the Hobart casino, a television ad and a visit by British environmentalist David Bellamy.

Luckily, Graham Richardson remained on the side of the conservationists. He said public sentiment would be the deciding factor in determining how much of the forests would be preserved. He brokered an interim deal he claimed would protect about 70 per cent of the forests considered by the inquiry.[21] In his opinion, 'every piece of available evidence showed that virtually the whole of the Lemonthyme and Southern Forests area would qualify for World Heritage listing'.[22] His package included the Lemonthyme and Farmhouse Creek forests, forests in the Wayatinah area near the Central Plateau as well as forests along the upper Weld, Gordon and South Styx rivers.

Richardson was lucky to get the backing of federal cabinet for his stance, nearly losing out to pro-development ministers.[23] In August, the Hawke government decided on slightly expanded protection and began pushing for an enlargement of Tasmania's World Heritage Area. Needless to say, Premier Gray was outraged and only agreed to go along with Canberra's wishes after extracting major concessions. He obtained an undertaking from the Hawke government for $52 million in compensation for the forgone forest resource, negotiated a huge increase in the state's volume of export woodchips and secured agreement that some major areas would be dropped from the federal government's proposed World Heritage Area expansion. Ominously, Gray also obtained an undertaking that the federal government would not pursue any more unilateral expansions of Tasmania's World Heritage Area, nor any more inquiries into the state's forestry industry.[24] The Hawke government was walking away from taking responsibility for protecting Tasmania's wilderness. Gray had extracted a heavy price.

The Wesley Vale pulpmill

While the Helsham Inquiry was being held, a seemingly innocuous article appeared in the north-west Tasmanian newspaper *The Advocate* in December 1987. It announced that North Broken Hill—owner of two of the state's three woodchip companies as well as the paper mill at Burnie—was proposing to build a major pulpmill in the farming district of Wesley Vale on the north coast in addition to a small mill already constructed there in 1961. North Broken Hill was to be joined in the venture by a Canadian company, Noranda. The article caught the attention of local teacher Christine Milne. Milne had been involved in the Franklin blockade as well as the fight for the Lemonthyme forests, and had recently organised a successful campaign to save some historic huts near Cradle Mountain. She and other locals were horrified by

North Broken Hill's plans. Their horror was heightened after they learnt the mill planned to employ a kraft pulping process that would use chlorine to bleach its pulp. This would result in large discharges of highly toxic and long-lasting dioxins into the nearby ocean. The mill would also significantly increase the pressure on the state's forests with a projected consumption of about 1.8 million tonnes of woodchips per annum, equal to about two-thirds the amount of woodchips being exported from the state each year.[25]

Milne contacted the Australian Conservation Foundation to see what it was going to do about North Broken Hill's pulpmill but was told it was too involved with the Helsham Inquiry and that she, and others, would have to take up the fight themselves. So join the fight they did. They soon established their own pressure group, Concerned Residents Opposing Pulpmill Siting or 'CROPS'.

For Milne and CROPS, battling the pulpmill was a baptism of fire. They knew little about environmental campaigning and the realities of taking on a corporate giant like North Broken Hill. Milne herself was challenged by the media, which she thought would wonder what 'a housewife and mother of two from Ulverstone [would] know about political realities'.[26]

Novices they might have been, but CROPS developed some important allies. Some of its biggest supporters were local, hitherto conservative, farmers. Even the fairly staid Tasmanian Farmers and Graziers' Association backed the fight against the mill, as did the Tasmanian Fishing Industry Council and some unions. The broad alliance owed much to the fact that the mill touched on many interconnected environmental issues. Milne explained: 'Wesley Vale was no single issue but demonstrated the interconnectedness of environmental concern. Encompassed in the debate were loss of native forests, appropriate land use, toxic pollution of air and waterways, depletion of greenhouse gases, contamination of food, recycling and waste, public health, community involvement in decision making and local self-determination.'[27]

The issue of appropriate pulpmill technology was central to the Wesley Vale debate. Milne and CROPS argued the kraft chlorine process proposed for the mill was substandard technology and that it should use cutting edge closed-loop technology instead that would release no ocean pollutants. Milne contended that North Broken Hill and Noranda were trying to build the Wesley Vale mill on the cheap. This was an important point: the conservation movement is often portrayed as being anti-development but the Wesley Vale issue was more about appropriate development. North Broken Hill and Noranda were so contemptuous of the appropriate development notion, however, that when asked by the state government to undertake an oceanographic study to determine the extent of the mill's chlorine pollution impact, they refused. They said the work 'will cost the company hundreds of thousands of dollars and cannot be justified until the project approval is secured'.[28]

Seemingly undaunted by the might of North Broken Hill and Noranda, Milne and CROPS organised comprehensive lobbying of local and national politicians. They also held a series of street rallies, attracting over 1000 people in Launceston at one rally in October 1988 and 12 000 in Hobart in February 1989.[29] Despite the opposition, Gray was unrelenting in his support for the mill. His government formally endorsed the proposal in October 1988. In February 1989 he infamously recalled parliament with a letter sent on North Broken Hill letterhead to pass legislation to weaken environmental benchmarks for the mill. A month later, the CSIRO (Commonwealth Scientific and Industrial Research Organisation) declared the benchmarks inadequate.

Although Graham Richardson had pledged to stop the mill if it could be clearly shown to be environmentally harmful, in March 1989 the Hawke government gave its support conditional on the mill's meeting tougher environmental guidelines than those required by the Tasmanian government. This support, however, was somewhat begrudging; even the pro-development minister for resources, John Kerin, said, 'the EIS is ratshit and the guidelines are laughable'.[30] But then, out of the blue, on the same day the federal government gave its conditional backing, Noranda announced it was pulling out. Its reasons remain unclear. Doubtless the company was worried by the opposition to the mill but some observers noted it was also concerned it could not meet the federal government's guidelines. More importantly, the requirements for the proposed mill could end up obliging Noranda to use better technology and meet higher standards in its own Canadian mills. A downturn in the global pulp market at the time was also likely to have played a part.

Suddenly the Wesley Vale issue was over. Milne and CROPS had scored a stunning victory. Premier Gray called an election to try to get the pulpmill back on track. Milne successfully stood as a 'Green Independent' (an early incarnation of the Greens in Tasmania) in the 1989 state election. North Broken Hill revived talk of the mill but the issue was put to rest when the Green Independents signed an Accord, an agreement with the minority Labor government that included a guarantee that the Wesley Vale mill would not go ahead. The Accord was a product of the Green Independents winning five seats and the balance of power in the state's lower house in the state election. They undertook to support the minority Labor government in return for conservation undertakings. The Green Independents had previously held two seats after a state election in 1986 and had begun their Tasmanian parliamentary presence when Bob Brown won a seat in 1983.

The Huon Forest Products mill

In mid 1986 another major Tasmanian forestry mill, this time a woodchip mill, was announced for Whale Point on the Huon River in south-east Tasmania. This mill was to be built by a consortium called Huon Forest Products (a group of sawmillers and the Australian Newsprint Mills company). Although its environmental impact would have rivalled Wesley Vale's, it was overshadowed by the northern pulpmill and never assumed the same profile. Like the Wesley Vale mill, the opposition to the Whale Point mill was spearheaded by a local pressure group: the Huon Protection Group. And just as for the Wesley Vale mill, the pressure group was headed by a woman—Peg Putt, a tireless campaigner who would go on to lead the Greens in the Tasmanian parliament (as Christine Milne would).

The Huon River is a majestic waterway that rises on the slopes of Mount Anne, the highest mountain in the south-west wilderness. It flows through the heart of Tasmania's wilderness and is flanked by extensive stands of forests. Any woodchip mill that threatened the Huon area would eventually pose a danger to the south-west at large. Putt and the Huon Protection Group did an extraordinary amount of grass roots campaigning against the mill, including letterboxing all of the Huon Valley and holding a series of local community meetings, one of which was disrupted by a local mob of disgruntled forestry industry sympathisers.[31]

Another prong to their campaign involved joining the Labor Party and—almost naively—taking a motion to its state conference in November 1988 calling on the party to oppose the mill. The Tasmanian Labor Party is the party of premiers Albert Ogilvie and Eric Reece, who were ardent backers of forestry (and of hydro development), and it was the party that allowed the woodchip industry into Tasmania in the early 1970s. The Tasmanian Labor Party has a long history of backing development of Tasmanian's natural resources—at virtually any cost—so at first blush Putt's motion seemed to have little chance of success. But a combination of extensive lobbying and alliance-building, in addition to Labor's hunger for power after a six-year absence from government, allowed Putt to win the day. TWS also assisted by lobbying sawmillers, the transport union and Labor Tasmanian senator John Devereux.

Like Milne and CROPS, Putt was new to the political game and like Milne she found herself being thrown into daunting new challenges by the mill campaign. As Milne and CROPS had done,

Putt and the Huon Protection Group latched on to the threat the mill posed to the water quality of the area as well as to the state's wilderness forests. The group came into possession of a leaked assessment by the state environment department that said the mill would destroy the already poor water quality of the Huon River. Thorough though their campaign was, Putt and the Huon Protection Group were not assured of victory until the signing of the governing deal between the Green Independents and the Labor Party that ruled out any future for the proposed Whale Point mill.

The Douglas–Apsley National Park

The Accord signed between the Green Independents and the Labor Party in May 1989 was a watershed in the fight for Tasmania's forests. Not only did it stop the Huon Forest Products mill and lock in the collapse of the Wesley Vale mill proposal but it also kept the longstanding state woodchipping limit of 2.889 million tonnes in place and, crucially, placed a moratorium on the logging of unreserved National Estate forests, including a cessation of logging at Jackeys Marsh. The Accord also set up a review that would investigate alternatives to the logging of National Estate forests (see appendix 3).[32] In the early Accord negotiations Bob Brown and the Green Independents had wanted a complete National Estate logging ban but Premier Michael Field, leader of the minority Labor government that won office in 1989, refused and they had to settle for the review instead.[33]

Another major undertaking in the Accord was one to make the forests of the east coast area around the Douglas and Apsley rivers a national park by the end of 1989. Unlike the wet temperate forests of the south-west, the forests of the Douglas–Apsley are mainly dry schlerophyll forests. By the late 1980s much of the state's eastern forests had been logged and the Douglas–Apsley area was the last unlogged catchment of forest still standing in that part of the state. There had been a campaign to save the Douglas–Apsley forests for some time. As early as the late 1970s there had been lobbying for their preservation and in 1984 a number of conservation groups put forward a proposal for a 14 800-hectare national park in the area which was backed up by tours of the area as part of the lobbying effort.[34] Unlike previous national park proposals this one had the support of most local businesses, which saw the new park as a tourist drawcard, a view that became increasingly common in the state. Support for the proposed park was enhanced by the work of local conservationists, who built tracks in the area and published maps of the region.

True to form, however, the Gray government was keen to log the area so it rejected the proposal and allowed road building to start in the Douglas–Apsley forests in 1986. The following year, however, the federal government put pressure on both Tasmanian Pulp and Forest Holdings—the company wanting to carry out the logging—and the Gray government to temporarily halt the forestry operations. It argued the operations would pre-empt the Helsham Inquiry, which was to include the Douglas–Apsley forests in its considerations.

The same year a similar tactic was used to that employed for the Huon Forest Products mill when a motion was successfully passed at a Tasmanian Labor Party state conference to protect the area's forests (partly as a result of lobbying by Phillip Hoysted from the Tasmanian Conservation Trust). The party included the promise in its manifesto for the 1989 state election.[35] The profile of the Douglas–Apsley forests was also lifted by the staging of a successful forests festival on the banks of the Apsley River in 1987.

Finally, in December 1989, a large group of people gathered on the banks of the Apsley to see the area formally declared a new national park after Labor delivered on its Accord undertaking. Forests that only three years before seemed doomed were safe, thanks to the persistence of a number of campaigners who never gave up on their vision.

The 1989 World Heritage Area expansion

Although the Douglas–Apsley National Park declaration was a great victory for the green movement, the really big conservation victory from the Accord was a major expansion of Tasmania's World Heritage Area. By the end of 1989, agreement was reached between the Green Independents and the Field government to double the size of the World Heritage Area to 1.38 million hectares or about 20 per cent of the state. The expansion was a major achievement for the five Green Independents, Bob Brown, Christine Milne, Gerry Bates, Dianne Hollister and Lance Armstrong. Together with his leadership of the Franklin campaign and his establishment of the Greens as a national and state political force, this will be one of the things Brown will be most remembered for.

The World Heritage Area expansion had several components. The Gray government, largely as a result of the Helsham Inquiry, had agreed to additions which amounted to 254 000 hectares. Included in this component were the Lemonthyme forests, forests around the Beech Creek area on the eastern boundary of the south-west, and forests around the Picton and Huon rivers on the south-east edge of the south-west wilderness. It also included the Walls of Jerusalem National Park. The negotiations between Brown and the Field government added another 317 000 hectares. Included in this were the western Central Plateau, Hartz Mountains National Park, the Little Fisher River, the Eldon Range, the Campbell and lower Gordon rivers as well as the Denison and Spires ranges. Another 34 000 hectares identified by the state and federal governments made up a third component, mainly comprising of areas near the west coast, while areas identified by the Salamanca Agreement process (see below), including forests on the Weld River and forests in the Tiger Range near the Florentine Valley, made up a fourth component.

The original Tasmanian World Heritage Area was a product of the Franklin campaign and its doubling hugely increased its environmental integrity. The expansion did not save all the contentious forests around the state's south-west wilderness area, however. At least two-thirds of the southern forests that formed the south-eastern flank of the south-west were left unprotected, as were most of the forests that ran down the northern escarpment of the Central Plateau.

The Salamanca Agreement and resource security legislation

Significant though the World Heritage Area expansion was, the part of the Green Independent–Labor Accord that held the greatest promise for the preservation of the state's forests was the commitment to set up a review into the feasibility of halting logging in all of Tasmania's National Estate forests, with a view to adding them to the state's World Heritage Area. Following through on this commitment, in August 1989 an agreement was signed between the forestry industry, the state government and the conservation movement—the Salamanca Agreement—to establish the review. The agreement was a high water mark in rare cooperation between all sides of the Tasmanian forestry debate. If successful, it could have removed the need for future forest battles and might have been a template for other forestry debates around the country. It was a bold experiment that could have made for wins all around.

For the conservation movement, the Salamanca Agreement represented another new framing of its forest campaign. The campaign moved beyond the World Heritage focus of the Helsham Inquiry to all of Tasmania's forests, including those on private land. Small regional forests were included in the movement's reserve proposals for the first time and its focus began turning towards the state's old-growth forests.

Throughout the first half of 1990, the Salamanca Agreement worked well and there was increasing confidence in its operation. But bit by bit the confidence ebbed away as the Field government put increased pressure on the process. Early on, the government unilaterally declared there would be no more World Heritage nominations without broad community agreement. Then the government said the mining industry should have a major voice in the agreement and could veto any new reserves. It then began imposing a cascade of further conditions designed to scuttle the process, signalling that it did not want the experiment to work. Despite the crumbling support, however, a draft strategy was released in June 1990 but within a fortnight Field made it clear he would not support its recommendations.[36]

From then on the process was on a slippery slope. Finally, in September 1990, the negotiations collapsed (after the government and unions used their numbers to outvote the conservation movement) and the Accord between the Green Independents and Labor also began to look shaky. The Green Independents increased the pressure by declaring 'proper management of the forests is more important to us than our seats in parliament'.[37]

Field had always been a major backer of Tasmania's forestry industry and once the forest talks started collapsing his government quickly moved to bury what remained of the Salamanca Agreement process as well as the Accord undertakings. In October 1990, his government increased the export woodchip quota above the ceiling agreed in the Accord. In November it began calling for expressions of interest in the construction of a pulpmill in northern Tasmania. Having destroyed the Salamanca Agreement, the government was tearing up the Accord. The most ominous moves made by the Field government, however, were approaches it began making to the federal government to interest it in passing 'resource security' legislation which would guarantee high levels of timber resource to the forestry industry. This legislation would complement state legislation. It represented a reframing of the forests debate by the industry around the supposed security of its resource. Resource security was the forestry industry's fightback against the gains the environment movement had made with the doubling of the World Heritage Area, the creation of the Douglas–Apsley National Park and the ditching of the Wesley Vale and Huon Forest Products mills. And, of course, it was about guaranteeing timber to the forestry industry and ignoring the conservation needs of Tasmania's forests.

Field was adamant about getting the resource security legislation through parliament despite the fact that the Green Independents made it clear they would never support it. By early 1991 his minority government and the Greens were well and truly deadlocked over the legislation. It became obvious the Greens were prepared to bring the government down over it. Brown, in particular, wanted to do more than just register a protest vote. He wanted to leave no doubt that the Greens could never support a government that backed such legislation. Field thought the Greens' main gripe was that the new legislation would break the Accord undertaking to limit woodchip exports to no more than 2.889 million tonnes per annum but their concerns were much broader than that.[38]

By May 1991 Field was putting forward a number of proposals to appease the Greens. None worked. In September he tried one last concession. There would be a ten-year logging moratorium on 600 000 hectares of key forests—'deferred forests'—if the legislation was allowed to pass (most of the forest was west coast rainforest that was not threatened by logging). The Greens found this proposal inadequate and continued their opposition. Throughout 1991 the tension grew as it became clear neither side would back down. TWS added to the pressure by letterboxing a leaflet throughout the state that proclaimed: 'It's not resource security: it's resource robbery'.

Finally, on 29 October 1991, Field introduced the resource security legislation in parliament and the Greens made good on their pledge to bring the government down by immediately introducing a no-confidence motion. Tensions had not run so high since the Franklin campaign. Facing the fall of his government,

Cartoon by Ron Tandberg that appeared in *The Age* in 1982 commenting on forestry in Tasmania.

Field withdrew the legislation and assured parliament he would not reintroduce it.[39]

Field seemed to have blinked in the face-off. Pandemonium broke out within the forestry industry. Amidst the confusion, Field then brokered a secret deal with the Liberal opposition that was facilitated by union boss and later pro-forestry premier Jim Bacon: the Liberals would join with Labor in passing the legislation in return for their being free to pass a no-confidence motion in the Field government after the end of the year. It was a bizarre deal; even though it gave Field a few more months, it still meant his government would fall. Clearly, he was more prepared to sacrifice his government than to walk away from the forestry industry. His support for the industry seemed completely unqualified.

With the support of the Liberals, the legislation was passed but the lopsided deal left Field with little choice but to call an early state election for February 1992. This he lost, with Labor reduced to eleven seats out of 35. The death of Field's minority government was also the death of any hope of cooperation in the Tasmanian forests debate. The opportunity for trust lasted less than three years. The Field government squandered it with its capitulation to the big forestry companies. What could have been a golden age in Tasmanian forestry politics ended in acrimony and was never revisited.

6

The Regional and Lennon–Howard forest agreements

The end of the Field Labor government in 1992 marked the end of consensus in the Tasmanian forests debate. All sides retreated to their trenches. The conservation movement returned to the forests of Farmhouse Creek in the Picton Valley. By then the gloves were well and truly off and the blockade was marred by the firebombing of two of the protesters' cars. In response to the blockade, the new state Liberal government of Premier Ray Groom passed legislation that broadened the powers of police to arrest people for trespass and increased the penalty for the offence. Groom made no secret of the fact it was aimed at the blockaders.[1] The Groom government's unequivocal support of the forestry industry was confirmed by its approval in 1993 of a new woodchip mill at Hampshire, in the state's north, in a large area of freehold forest owned by Associated Pulp and Paper Mills (APPM).

Groom made it clear there would be no more forest conservation wins under his government. Alec Marr, from The Wilderness Society (TWS, its name changed from the Tasmanian Wilderness Society in 1983), observed: 'it was quite clear we could not solve the Tasmanian problem inside Tasmania with the set of dynamics that existed post-accord'.[2] The attention of the conservation movement turned to Canberra, where it engaged in an intense lobbying campaign in an attempt to limit the granting of federal woodchip licences at the end of 1994. The Keating government's environment minister,

John Faulkner, was sympathetic to the conservation movement (though less so than Graham Richardson had been). But the federal resources minister, David Beddall, had ultimate responsibility for the annual issuing of woodchip export licences. Beddall was not sympathetic to conservation, once even describing trees as 'vertical logs'. The conservation movement pressured Faulkner and Beddall to make the woodchip licence decision consistent with the National Forest Policy Statement that was signed by all states and territories, apart from Tasmania, in December 1992. The statement sought to conserve old-growth forests and wilderness and aimed to create a system of forest reserves (set-aside forest conservation areas) throughout the country by 1995, although little progress had been made in this direction by the time the woodchip licence decision came around.[3] Faulkner visited the forests of the Picton and Huon areas in April 1994, accompanied by Bob Brown and Geoff Law, and was comprehensively lobbied by Alec Marr from TWS. Faulkner recommended to Beddall that 1311 key sensitive forest coupes (forests earmarked for logging) around Australia be excluded from the renewed licences, including about 150 in Tasmania. But Beddall ignored the advice and renewed the licences for nearly all of the 1311 areas. Keating endorsed Beddall's decision. Commenting on Keating's support, Bob Brown said: '[he has] a better feeling for the Japanese woodchip contract than for the koala habitat'.[4]

There was a strong reaction against Beddall's blatantly pro-forestry decision: 80 000 people attended forest rallies around the country in early 1995, including over 5000 in Hobart. The Greens called for Beddall to be sacked. The backlash was reinforced by the success of a court action the Tasmanian Conservation Trust brought against a woodchip licence issued to forest products company Gunns Ltd, the granting of which, the Trust argued, had not followed proper procedure. Keating could see the backlash becoming a major liability so in early January he stepped into the fray, giving interim protection to 72 coupes in Tasmania that formed part of a nationwide package of 509 coupes in which logging would be suspended pending assessment of their environmental values.[5]

But in attempting to relieve one pressure, Keating created another—Groom pledged to log the protected coupes, regardless of the federal government. In response to the Prime Minister's intervention, at the end of January 350 log trucks descended on Parliament House in Canberra to blockade it. The trucks ringed Parliament House and politicians entering the building had to run a gauntlet of noisy drivers. Keating backed down; throughout February and March more than half of the Tasmanian coupes he had earmarked for protection were released for logging. His government then began talks with state governments on long-term federal–state forestry management agreements, perhaps in the belief that the process would somehow take woodchip licences out of the headlines. This was a major backwards step for the forests and was part of what the environment movement generally saw as a Labor government retreat from environmental issues under Keating's leadership. The talks led to the creation of the Regional Forest Agreement (RFA) process, which was continued after the election of the Howard government in March 1996, a government that proved to be still weaker on environmental issues than the Keating government.

Shortly after its election, the Howard government made its attitude to forestry clear by significantly increasing the maximum allowable tonnage of Tasmania's woodchip exports. In Tasmania, however, the Greens held the balance of power in state parliament between 1996 and 1998. Greens parliamentary leader Christine Milne applied a lot of behind-the-scenes pressure to the government of Premier Tony Rundle, which softened some of the edges of the Howard government's position. TWS also held a major forests rally outside the Tasmanian parliament in September 1997. By then there was no doubt about the enormity of the woodchipping threat to Tasmania's forests. Over 80 per cent of all logs taken from the state's native forests were going to woodchip mills. The Howard government's decision to increase the woodchip quota created major concerns amongst the conservation movement about what was planned for Tasmania's RFA. These fears were confirmed when the draft agreement was unveiled in 1997.

On paper, the RFA seemed almost green: its aim was to preserve at least 15 per cent of the forests that existed at the time of European settlement of Tasmania. It was also designed to conserve at least 60 per cent of the 1996 area of each major type of old-growth forest, including at least 90 per cent of wilderness forests. The RFA called itself a 'comprehensive, adequate and representative' forest reserve system. But the devil was in its detail. The RFA used a narrow definition of old-growth forests; it used a broad definition of already protected forests; it did not impose any moratorium on logging in sensitive forests while it was being negotiated; it allowed for little genuine public input; and it was slanted by a number of biased social and economic studies. It also had a get-out clause that allowed key forests to be excluded from reservation if there were major socioeconomic reasons for doing so.

The RFA was meant to be a long-term answer to Tasmania's forestry divisions but quickly showed that it would resolve nothing because it was firmly on the side of the forestry industry. In the end the RFA only reserved 70 per cent of the state's wilderness forests[6] and only 48 per cent of its magnificent tall old-growth *Eucalyptus regnans*.[7] Despite the agreement's ambitions to reserve significant areas of remaining old-growth forest, it ended reserving modest new areas of forest and abolished the already high ceiling on the state's woodchip exports. This led to large increases in the state's woodchip cut, despite the RFA being committed to 'ecologically sustainable forest management'. By far the worst element of the package was its giving of $76 million to the state forestry industry that was used to fund the conversion of native forest to plantations. This ended up destroying nearly 100 000 hectares of mostly tall eucalypt forest in the north and east of the state which would have been uneconomic without the subsidy.

There were some wins from the RFA for the environment movement, however. Some small stands of forest along the eastern boundary of the World Heritage Area—including along the Huon, Picton and Counsel rivers—were protected, as were some forests along the northern edge of the Central Plateau that were placed above logging boundaries. A crescent of forest reserves was created in the east between Buckland and Bicheno. The RFA also led to expansion of the Freycinet and Mount William national parks as well as the creation of the Savage River and Tasman national parks. But considering it was meant to strike a long-term balance between the interests of the industry and those of conservation, it was disappointing. When Prime Minister Howard and new Premier Tony Rundle signed the RFA in November 1997 they were met with a noisy demonstration. Geoff Law simply called the agreement 'a disaster for the forests'.[8] Activists entered state parliament and dropped a giant banner over the front of the building that read 'Rundle's Forest Atrocity'. The Greens considered bringing the Rundle government down over the RFA by putting up a no-confidence motion, as they had with the Field government and its 1992 resource security legislation, but decided against it because the Bacon-led Labor opposition had a worse position than the Liberals on forest protection. Brown felt that a major cause of the inadequate forest reservation in the RFA was pro-forestry pressure from the state Labor opposition and the unions. He observed that '... any moves towards protecting forests under the RFA were constantly criticised by Labor as going too far. It was pressure from the Labor opposition in Tasmania, the TTLC [the Tasmanian Trades and Labour Council] and the timber industry which prevented the Liberals from having room to move to protect forests.'[9]

The Mother Cummings post-RFA protest

It was not long before the credibility of the RFA was put to the test. No sooner was the ink dry on the agreement than loggers moved into a key coupe in the Mother Cummings Peak area of the Great Western Tiers. The coupe had a large stand of sub-alpine tall forest, mainly made up of old-growth *Eucalyptus delegatensis* with a rainforest understorey. During the RFA negotiations it had been given a high priority by the conservation movement but ended up without protection in the final agreement. The coupe also had a special significance for the local Aboriginal community, who called the Great Western Tiers area *Kooparoona Niara* (mountains of the spirits). So when loggers moved into the Mother Cummings area it became the site of a major post-RFA showdown. The Mother Cummings blockade became the largest forest direct action Tasmania had seen since the Farmhouse Creek and Lemonthyme blockades. At its peak in March 1998, 1000 people protested at the site with 60 arrested on one day alone. The icon of the blockade became Neil 'Hector' Smith, a 52-year-old computer programmer who set up a sophisticated platform 25 metres up a tree in the path of a proposed logging road. Smith became known as 'Hector the Protector' and literally took the art of tree-sitting to a new level. He included a computer on his platform, from which he sent information about the blockade to the world. As had happened at the Farmhouse Creek blockade, loggers tried intimidating Smith by felling trees next to his platform. After Smith had occupied his platform for thirteen days the police finally climbed up and arrested him.

The Mother Cummings blockade went on for six weeks and tied up 40 police, costing the state government more than twice the royalties it would get from logging the area. With this blockade both the police and the state government showed no mercy. After numerous court hearings, Smith was eventually fined $7000 for 'interfering with the operations of a vehicle'. Fellow blockader Sarah Bayne was fined $5000. Another protester, Karen Weldrick, was fined $3000 for locking herself to a log truck, for which she eventually spent 31 days in gaol after being unable to pay the fine. The blockade received sympathetic media coverage and was supported by numerous rallies around the state. Although the industry and state government were determined the RFA would end Tasmania's forests clashes, the Mother Cummings blockade signalled in no uncertain terms that the debate had not been quelled.

In August 1998 an early state election was held and after six years in opposition the Labor Party was returned to office under its leader, Jim Bacon. Bacon had helped negotiate the passing of the 1992 Field government forestry resource security legislation and was every bit as supportive of the forestry industry as previous Labor and Liberal premiers had been. Within months of his election he gave effect to the RFA. He also gave the green light to the logging of the deferred forests identified by the Field government. These forests included parts of the Tarkine rainforests in the north-west; forests in the Weld, Styx, Counsel and Florentine valleys on the eastern side of the south-west wilderness; Western Bluff in the Great Western Tiers; parts of the forests in the north-east highlands; and rainforests on the west coast.[10]

The Tarkine road

One of the few major gains the conservation movement got from the RFA was the creation of the Savage River National Park in the state's north-west. The park protected some of the extensive rainforests in that part of the state known as 'the Tarkine'. The name, which comes from the Aboriginal people who once occupied the area, was suggested by Bob Brown. What made the Tarkine unique was that it was home to the largest tract of temperate rainforest anywhere in Australia. Apart from some mining activity, the Tarkine had escaped major development and is Tasmania's second great wilderness area after its large south-west wilderness.

By the time the RFA was signed, the Tarkine was well known. The issue that brought it to prominence was the building of a road through the area. The idea of such a road had been around for a long time and in 1987 work commenced on the project but was stopped in 1989 by state Labor transport minister Ken Wriedt after lobbying from TWS. By 1992 the prospect of a recommencement of work was being taken seriously, prompting the formation of a Tarkine Action Group in Launceston.[11] In 1994 work on the road resumed without any environmental, economic or social study being conducted into its impact. The road was in an area that was seldom visited and which already had good vehicular access at either end; it was also to pass through challenging country. It seemed a senseless development and soon earned the tag of the 'Road to Nowhere'. A group of young, committed activists—who became known as the 'Tarkine Tigers'—was determined to stop the road. They set up a base near Burnie and persistently blockaded work on the road throughout 1995. Along with other local activists they helped make the Tarkine a household name, enduring harsh conditions, arrest, harassment by workers and even gaoling. In the end, the Tigers were unsuccessful and the 53-kilometre road opened in January 1996.

The road controversy was one of a number of campaign fronts that raised the profile of the Tarkine throughout the 1990s. TWS proposed the area for World Heritage listing in 1992 and had it listed on the Register of the National Estate. TWS also produced posters of the Tarkine and published a book of photographs from the region in 1995. It also became part of a national Tarkine coalition of environment groups that lobbied for protection of the area. The Tarkine was a constant focus of the forest lobbying carried on by TWS, which emphasised the value of precious rainforests of the area.

Throughout the 1990s the focus of the Tasmanian forests campaign turned away from National Estate forests to concern for the fate of old-growth forests. Although these forests embraced much the same areas as the National Estate forests, the new label was self-explanatory and more evocative. And there was no doubt that old-growth forests remained very much under threat. Although the Tasmanian forestry industry often presented itself as selective in its logging of old-growth forest, in 1999–2000 two-thirds of all the native forest logged on public land was old-growth.[12] Only a quarter of the old-growth forests that existed when Europeans colonised Tasmania still remained. In 2001–02 alone, 32 000 hectares of native forest were approved for logging.[13] The amount of native forest that was logged then replaced by plantations was a contributing factor in land clearance being a bigger problem, in relative terms, in Tasmania than in any other state (including Queensland).

Such enormous areas of forest can be logged because the forestry industry controls large slices of the island. Forestry Tasmania (since 1994 the name of the former Forestry Commission of Tasmania) is currently responsible for 1.5 million hectares of forested land, equal to 22 per cent of the state,[14] while forests on private land cover an additional 1 million hectares. After allowing for forest reserves, which cover 3 per cent of the state, public and private commercial forests covered 37 per cent of Tasmania. Tall trees have borne the brunt of forestry clearance: just 18 per cent of the tall trees that existed at the time of European colonisation remain as old-growth and only 54 per cent of that vestige is adequately reserved.[15] Chilling statistics such as these, in tandem with the old-growth forests and tall trees messages of the conservation movement, began resonating with the Tasmanian public by the later part of the 1990s. It sensed something was seriously wrong in Tasmania's forests.

When Tasmanians were polled in 2001 about whether they supported the end of clearfelling within the next 10 years, 69 per cent said they agreed with the idea and only 21 per cent said they were opposed. Similarly, when asked whether they supported protection of tall old-growth forests, 70 per cent said they would like to see such trees permanently protected.[16] A similar message came through in a national poll commissioned by the Doctors for Forests group in January 2004, which showed 85 per cent of people across the country supported federal government intervention to stop the logging of Tasmania's old-growth forests.[17]

The new plantation threat

By the late 1990s it was clear that forest plantations were posing a large and growing threat to the state's old-growth forests. Although once regarded by the environment movement as a panacea to the logging of native forests, it was becoming obvious that plantations were adding to the pressure on Tasmania's native forests, not reducing it.

In the mid 1990s the federal government began a scheme that extended generous tax concessions to investors in new forest plantations. It also pumped $76 million into the Tasmanian forestry industry after the RFA, much of which was used to subsidise plantations. The result was an explosion in plantation establishment in Tasmania; more than a quarter of a million hectares by 2002.[18] The plantations were established either on farmland or in areas where native forest had been clearfelled. About 15 per cent of the state's farmland was taken over by plantations. Often the farmland was prime agricultural land. Worse than this was the fact that by 2000–01 nearly 8000 hectares of native forest was being cleared for plantations each year.[19] Between 1999 and 2006, 64 000 hectares of native forest in Tasmania was converted to plantations.[20] More than 60 per cent of all native forest clearfelled in the state was making way for monoculture tree farms.[21] Plantation establishment came to drive much of the native forest logging effort and was turning into a major environmental hazard. Compounding its menace was the fact that a poisonous herbicide—atrazine—was used to stop other vegetation from competing with plantation seedlings. The poison worked its way into nearby waterways, causing major health liabilities for people and animals.

The environmental problems of plantations were further compounded by the fact that Tasmania was the only state to use a poison called 1080 to stop native animals from grazing on the seedlings. Although effective in warding off animals, it resulted in the deaths of hundreds of thousands of native animals including wallabies, possums, bettongs, wombats, pademelons and potoroos.[22] The odium of the conversion of native forest to plantations grew to such an extent that the 2005 Lennon–Howard forest agreement stipulated that the practice had to end by 2010 on public land and 2015 on private land, a deadline that was eventually brought forward to 2007 (though only Gunns and Forestry Tasmania have so far agreed to this new date and there are several loopholes in the new regime).

The underselling of Tasmania's forests

If plantations were a relatively new development of Tasmanian forestry, a long established feature was the underselling of its timber. Getting too little return for valuable natural resources is a common theme in Tasmania's hydro, mining and forestry sectors. Big pulp, sawlog, metal and manufacturing industries have long received a bargain in the state. It has been known for some time that Tasmania was getting too little for its timber. In 1977 a state government inquiry headed by former deputy premier Merv Everett found that royalties charged by the Forestry Commission were 'quite uneconomic', a conclusion echoed by a 1972 upper house inquiry.[23] The Forestry Commission of Tasmania, and subsequently Forestry Tasmania, have often run at a loss or have made only modest surpluses. In 1989 forest economist Randal O'Toole noted: 'The Forestry Commission has cost Tasmanian taxpayers nearly $660 million (in 1988 dollars) over the past 50 years. This is nearly $1500 for each resident of Tasmania.'[24]

In recent years the forest service has managed to make some surpluses but they have been modest: between 2001–01 and 2004–05 Forestry Tasmania made an average surplus equal to less than a 1 per cent return on its net assets. If Forestry Tasmania had not had several debts forgiven, and not been given several government handouts, it would be making constant losses. If it was a regular business it would have gone bankrupt long ago. In 1990 Forestry Tasmania had $272 million of accumulated debt forgiven by the state government when it was corporatised and in 1998 $52 million of softwood plantation loans were forgiven by the federal government. In addition it has received a large share of $52 million given to the Tasmanian forestry industry after the 1988 Helsham Inquiry, $76 million given after the 1997 Regional Forest Agreement and $235 million extended after the 2005 Community Forests Agreement. But still it struggles to make money.

Forestry Tasmania has recently made it clear that it sets royalties on a sliding scale according to market timber prices. It wants observers to believe that it takes part in a fair sharing of profits with the state's forestry companies. But the reality is that the state's timber companies are making large profits while Forestry Tasmania struggles. A fair profit-sharing is not taking place at all. Between 2000–01 and 2004–05, while Forestry Tasmania's return on net assets averaged less than 1 per cent, timber company Gunns' equivalent return averaged 19 per cent.[25] Gunns was making super profits while Forestry Tasmania made next to nothing. Victoria has recently gone in the opposite direction to Tasmania by setting up auctions for its hardwood timber. Tasmania should do the same instead of selling timber at rock-bottom prices.

The Tasmanian forestry industry likes to talk about the employment it is responsible for creating but that, too, is poor compensation for the low royalties being charged. The largest number of jobs there have ever been in the Tasmanian forestry industry was 11 200 in 1984. Despite a significant increase in the amount of low-grade eucalypt cut since then, job totals today sit at around 6500, slightly more than half the level of two decades ago. According to the pro-conservation lobby group Timber Workers for Forests, the falling labour intensity of the state forestry industry is reflected in the number of jobs created from every hectare of forest it clearfells. In 1994–95, 1.3 forestry jobs were created for every hectare clearfelled but by 2003–04 this had fallen to just

0.35 jobs. However you view it, Tasmania is getting a raw deal on the sale of its forests.[26]

The Styx campaign

Talking about tall trees and old-growth forests is one thing, campaigning against threats to particular forests is another. Forest campaigns are invariably most successful when they make the community aware of threats to specific forests. The campaigns against the RFA process of the mid 1990s, as well as the woodchip EIS process of the early 1980s, showed the public did not respond to issues of forest management. Campaigns such as those against the logging of the Farmhouse Creek and Mother Cummings forests, however, showed the public did take up on particular threats. After the RFA, there were a number of threatened forests in the north of Tasmania that were easy for the public to access and campaign for, but no such forests in the south had been publicised by the movement. In 1999, however, TWS began publicising the tall old-growth forests of the Styx Valley, some 70 kilometres from Hobart, an extensive stand of old-growth forest that encapsulated for the public the grandeur of ancient, undisturbed forests.

The Styx is on the eastern border of the south-west wilderness area near the Florentine Valley that was once home to the largest stands of tall trees in Tasmania. The Florentine had been the focal point of the state's first conservation campaign in the 1940s (see Chapter 4). The Styx is home to many tall trees and is one of Australia's last strongholds of towering *Eucalyptus regnans*; a report released by Forestry Tasmania in 2000 noted that it contained the world's tallest hardwood trees. Fittingly, TWS called the area 'The Valley of the Giants'. Despite the grandeur of the Styx, in November 2000 Forestry Tasmania released plans to log over 800 hectares of the valley's old-growth forests over the following three years.

TWS publicised the Styx campaign in imaginative style. In December 1999 it organised a group of climbers and cavers to drape 3000 Christmas lights on a tall tree in the valley. The climbers spent eight days spreading the lights throughout the 80-metre tree, breaking the record for the world's tallest Christmas tree (although the record was disallowed because the tree was not a spruce). TWS continued its innovative campaign approach, holding church services in a large hollowed tree, arranging visits by prominent actors and musicians, scheduling bus tours to the area and producing a pamphlet that took visitors on a self-drive tour through the area's threatened forests. Within three years, over 1000 people had been on the tour. A five-month blockade was also staged in the Styx that included a high-tech tree platform similar to Smith's in the Mother Cummings blockade.

TWS's Styx campaign helped to significantly re-engage the public in its forest campaigns in the late 1990s. It started a new momentum that included well-attended Tasmanian rallies in July 2000, March 2001, August 2001, April 2003 and March 2004, the latter attracting 15 000 people. There were also major Tasmanian forest rallies in other states. In July 2003, in the middle of a Tasmanian winter, 3000 people travelled to the valley to voice their support for the preservation of its forests. TWS made the Styx a household name, ensuring that it loomed large in the 2005 Community Forest Agreement.

The 2004 federal election and Lennon–Howard Forest Agreement

The combined impact of the ongoing clearfelling of native forest, and of plantations, ensured that forestry remained a major issue in Tasmania well after the signing of the RFA. It played a prominent part in a state election held in 2002 and also had a high profile in the federal election of 2004. In the 2002 state election a conservation group, the Tasmanian Community Alliance, raised a significant amount of money which it used to air professional television advertisements. This helped make the plight of old-growth forests a major issue in that election.

An early signal that Tasmania's forests would figure in the 2004 federal election came in March of that year when Labor opposition leader Mark Latham spent two days in Tasmania talking to the forestry industry and doing a tour of the Styx with Bob Brown. After the tour Latham commented that he saw no reason to re-negotiate the RFA but also identified a phase-out of old-growth clearfelling in Tasmania as a 'very, very important priority'.[27] In late August, Prime Minister John Howard called a federal election for early October and within a week heightened interest in Tasmania's forests by saying: 'I think everybody would like to see old-growth logging stopped'. Ominously, though, he added, 'I do not support throwing regional communities on the scrap heap in order to achieve a particular environmental objective'.[28]

Howard's comments and Latham's visit ensured Tasmania's forests were a big election issue. The details of the forests policies from both sides were eagerly awaited. But neither wanted to show his hand early and both left their Tasmanian forestry announcements until the last week of the election campaign (despite advice from Brown to Latham).

When they did show their cards, Latham's package was far superior to Howard's because it reserved more forest and would result in more industry restructuring. Latham committed to the preservation of 240 000 hectares of high conservation value public forests (as well as another 165 000 hectares of forest on private land) pending review by a panel of experts, and pledged $800 million in restructuring funds for the forestry industry as well as new conservation incentives for private landowners. Howard promised to preserve 127 000 hectares of public forest (as well as another 43 000 hectares of forest on private land)—not necessarily of high conservation value—and only offered $52 million in compensation. He also said no forestry industry jobs would be lost. He distanced himself from any further federal interventions, declaring, 'My government does not intend to have any further inquiries into this issue'.[29]

Just days before the election, Howard was cheered by 1400 forestry industry workers at a rally in Launceston. By selling out on the forests, he managed to outmanoeuvre Latham and made his forestry package a means through which he could appeal to swinging working-class voters. Predictably, the forestry industry swung behind Howard and distanced itself from Latham. It placed full-page advertisements in Tasmania's papers with slogans such as: 'John Howard will protect Tasmania's forests and forestry workers, Mark Latham will lock up our forests and destroy thousands of jobs'. The state Labor government also turned on Latham. Premier Paul Lennon told forestry workers that Latham had 'bargained away' their jobs. TWS tried to balance the pressure by running a comprehensive Tasmanian forests campaign in marginal seats throughout the country during the election campaign.

Howard went on to easily beat Latham in the election and then began secret negotiations with the Tasmanian government to give effect to his Tasmanian forestry commitment. When the package was unveiled in May 2005 it was underwhelming. It claimed to have resulted in the reservation of an extra 148 000 hectares of forest on public land but, in fact, all but 58 184 hectares of the total was forest protected in informal reserves that lacked legislative backing and had little security of tenure. Most of the informal reserves were forests identified since the RFA as being in inaccessible areas that would therefore never be logged. About 45 000 hectares of the Tarkine's rainforest was reserved in the agreement but only about 5 per cent of all the highly productive forests proposed for protection by TWS outside of the Tarkine was protected.[30] In the Styx, 4671 hectares was reserved but this was equal to less than 40 per cent of the area proposed for reservation by TWS.[31] Unfortunately the 2005 saving of the Tarkine's rainforests led to a proposal by

Forestry Tasmania in 2007 that a 132 kilometre tourist road be built through the region, including 8 kilometres through rainforest.

Ironically, the secretly negotiated agreement was called the Tasmanian Community Forest Agreement. Key forests in areas such as the Weld, Huon, Florentine, Blue Tier, Leven Canyon, Reedy Marsh–Dazzler Range, the Great Western Tiers, north-east Highlands, Ben Lomond and the southern part of the Styx received virtually no protection. After Lennon had earlier pledged to stop the use of 1080 poison in all forests, the package allowed the poison's continued use on private land. The agreement also allowed conversion of native forest to plantation to continue for another five years on public land and ten years on private land (this was eventually shortened to finish in some forests in 2007). It declared that the clearfelling of old-growth forests could continue but would be reduced to about 20 per cent of all old-growth logging on public land by 2010. It unashamedly appeased the forestry industry by offering it $235 million in compensation: five times what Howard had originally offered. Although the package included some forest conservation gains, these were much smaller than Howard had promised. TWS called the package 'mediocre' and gave it four out of ten[32] while Brown said it was 'two parts poison to one part champagne'.[33]

The 2005 Lennon–Howard forest agreement was a supplement to the 1997 Regional Forest Agreement. The reaction to the agreement was more muted than the response to the RFA had been (partly because of the increasing need to focus on Gunns' proposed pulpmill; see Chapter 7) although a number of local forest campaigns did spring up or developed more momentum. A campaign launched in 2003 against the logging of key forests in Blue Tier, in the north-east, developed an increasingly high profile. A legal case brought by a group called 'Save our Sisters' (SOS) before the Lennon–Howard deal against logging in a key coupe on the east coast of the state also drew the attention of many (though eventually the group had to withdraw because of mounting legal

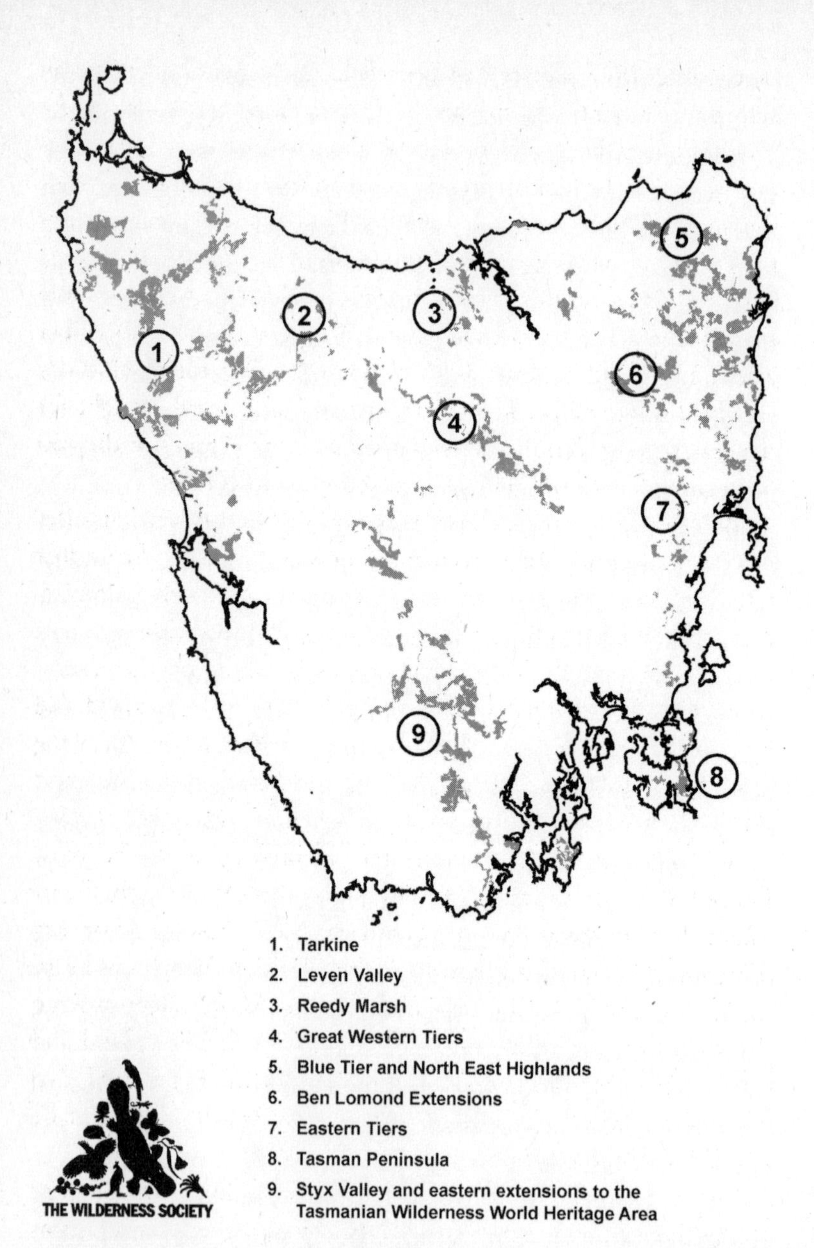

Figure 6.1 The significant unreserved forests of Tasmania in the wake of the 2005 Lennon–Howard Community Forest Agreement as identified by The Wilderness Society

costs). In 2006 a protest was again held in the Styx Valley and an action was mounted in the nearby Upper Florentine Valley where Howard's deal had only protected about a quarter of the old-growth forests he had originally promised to protect there. One of the most determined post Lennon–Howard agreement actions took place in the Weld Valley, on the eastern side of the south-west wilderness area, where activists staged a blockade that ran for more than twelve months between September 2005 and November 2006. The Weld action was the longest-running forest blockade ever held in Tasmania. The group that had successfully fought for the Tarkine's protection also continued its campaigning for the area by pushing for World Heritage protection of its forests.

Bob Brown spearheaded two other major forest campaigns after the Community Forest Agreement. One was a bold legal case that tested whether the RFA protected threatened species in forests that were earmarked by the forestry industry in the Wielangta area in the south-east. Brown was courageous in taking on the case; it could have bankrupted him. Luckily his exposure was reduced by a significant amount of fundraising. In December 2006 the judge hearing the case said the logging had breached the RFA and was likely to have a 'significant impact' on three endangered species. Brown called the win 'monumental', identifying it as the most important environmental test case since the 1983 High Court Franklin Dam decision.[34] In February 2007, however, Forestry Tasmania announced it would appeal the decision (which was heard in August 2007) and in December it won the appeal prompting Brown to petition to appeal to the High Court. The federal and state governments also announced they would change the RFA to redefine the words 'to protect' so that Brown's win could not be applied to other forests.

Brown also led the campaign to save forest on private land at Recherche Bay, on Tasmania's south-east coast, where French ships under the command of Bruni d'Entrecasteaux had landed in 1792 and 1793. The private owners of the Recherche Bay forests were

Cartoon by Ron Tandberg that appeared in *The Age* and *Sydney Morning Herald* in 1994 commenting on Keating's release of hitherto protected forestry coupes for logging.

determined to log the forest and the state government would only commit to preserving its fringes. But in February 2006 it was announced that Brown had brokered a unique deal with businessman Dick Smith, whereby Smith would help finance the purchase of the 140-hectare site. It was an innovative win that would not have happened without Brown, or the significant campaigning by local activists.

The 2004 election came twenty years after the Tasmanian environment movement first became seriously involved in the forests debate. Those twenty years were a formative time, demonstrating forestry's powerful and entrenched position in the state and teaching environmental campaigners to make the most of political opportunities when they arose. By 2004 the movement had a mature approach to forestry and probably thought it had seen the worst of what the industry could throw at it. But ensuing years would raise the forestry stakes even further and severely test the mettle of the movement.

7

Gunns

In 2000 Gunns, a company traditionally known in Tasmania for its construction business and hardware stores, bought one of the state's woodchip companies. When it moved in the following year to buy the remainder of the state's export woodchip industry, Alistair Graham, from the Tasmanian Conservation Trust, unsuccessfully tried to stop the concentration from going ahead. Unbeknownst to the environment movement, however, this was the start of a David-and-Goliath struggle where it often seemed only one side could be left standing.

The rise of Gunns

Throughout the 1980s and 1990s, the Tasmanian conservation movement did battle with two major export woodchip companies: Associated Pulp and Paper Mills (which owned Tasmanian Pulp and Forest Holdings and which was eventually bought out by Boral) and North Forest Products. By 2001, however, both these woodchipping interests had been gobbled up by Gunns. Originally established as a building business in 1875 by brothers John and Thomas Gunn, the company purchased local brickworks and sawmills to guarantee its supply of materials for its building activity.[1] Construction remained the company's main focus even after APPM bought a 30 per cent shareholding in the company in 1971. That focus changed in the 1980s, however, when a consortium of northern Tasmanian businesspeople bought out APPM's share. In 1986, the company listed on the stock exchange and the Gunn family divested its remaining interest. The company then began a series of major forestry industry acquisitions. In 1986 it joined forces with major sawmilling company Kilndried Hardwoods. It then bought the timber interests of the Kemp and Denning company in 1989, the Kauri Timber Company in 1993, Tasmanian Veneers in 1994, French's Pine World in 1994 and Hiddings Hardware in 1995.[2] In 1995 Gunns secured its first export woodchip licence and then in 1996 the one-time ardent backer of the Franklin dam, ex-premier Robin Gray, joined its board.[3] By the late 1990s the company was ready to expand even further, which it did in 2000 when it bought Boral's woodchip operations. Gunns consummated its dominance of Tasmania's forestry industry in 2001 with the purchase of the woodchip operations of North Forest Products. There was now one massive forestry company in the state that could intimidate the conservation movement. And intimidate the movement it did.

One of the northern Tasmanian businesspeople who controlled Gunns in the late 1980s was Edmund Rouse, chairman of Gunns Kilndried. Rouse was a colourful and powerful figure who was not afraid to throw his weight around the island's small community. By 1989 he had sufficient forestry interests to feel threatened by the possibility that the Green Independents might hold the balance of power after state elections to be held that year. Rouse told confidants that if that happened, he could lose up to $15 million.[4] When the Greens did win the balance of power, Rouse set about attempting to bribe a Labor member of parliament to vote with the defeated Liberal Party, which would have delivered it a governing majority. Rouse's move was audacious and showed blatant disregard for every democratic convention there is. After the police were alerted, Rouse was arrested and eventually spent eighteen months in prison. The Chief Commissioner of a Royal Commission conducted into the incident, William Carter, was damning of Rouse's conduct, saying: 'no community with a democratic system of government can tolerate or seek to excuse the type of crime Rouse committed'.[5] After serving his term in prison, Rouse began selling his business interests in Tasmania, including his forestry investments, and eventually moved to Melbourne.

It is hard to understate the significance of the company Rouse had a financial interest in. Gunns employs about 1700 people and is the biggest private sector employer in the state. It owns a string of timber and hardware stores throughout the island, several veneer and sawmills, a construction company, and two wine companies. After it bought all of the state's export woodchip operations its Tasmanian woodchip exports equalled one-fifth of all the hardwood woodchips traded around the world, making it the largest hardwood export woodchipper there was at the time. Gunns was well aware of its dominance and wasted no time in using it.

The 'Gunns 20' legal action

It was soon clear that Gunns was keen to use its new corporate might to directly affect the political processes of Tasmania. It became a major donor to the state's political parties, giving $50 000 to the Labor party and $25 000 to the Liberal party in 2002–03 alone.[6] Its political coup de grâce, however, came in December 2004, just after the bruising federal election in which Tasmania's forests were a major issue. On the morning of 14 December 2004, two months after the election and without warning, twenty people and organisations in the environment movement were served with legal writs from the company. Gunns was suing the group—which quickly became known as the 'Gunns 20'—for a staggering $6.3 million, $3.5 million of which was claimed from The Wilderness Society (TWS) and five of its staff. Also served with writs were Bob Brown, Peg Putt (state parliamentary Greens leader), the Huon Valley Environment Centre,

Doctors for Forests as well as a number of individual activists. Gunns alleged that various protests against its forestry operations—such as those in the Styx and at its woodchip mill at Hampshire—were illegal attempts to interfere with its trade. It also boldly claimed that various campaigns mounted by members of the Gunns 20 amounted to a conspiracy to injure it by unlawful means. It was a daring attempt to silence the environment movement by using a 'SLAPP' (Strategic Litigation Against Public Participation) style of writ that was popular with United States companies where a number of companies had used it to their advantage.

The Gunns 20 group was defiant. One of its defendants, Burnie dentist Peter Pullinger, declared: 'We will continue to defend Tasmania's ancient forests. We will continue to defend our clean air and water. We will continue to defend public health and to speak out in the interests of the Tasmanian community.'[7] The action raised the temperature of the Tasmanian forests debate to an unprecedented degree. It had the potential to wipe out the life savings of many activists as well as most of TWS's assets. One of the activists involved in the suit, Adam Burling, said it had put 'stress on me and my family' and had the potential for him to 'lose everything I've got'.[8] While the stakes were high for the conservation movement, they were also high for Gunns. The company could overreach itself in its action, making it look mendacious.

By the start of 2008 the writ was still being played out in the courts which, to date, have found more in favour of the conservation movement than of Gunns. Early on, it became evident that the writ was not well constructed. When it was first heard by the Victorian Supreme Court in July 2005 Justice Bongiorno called it 'unintelligable' and 'embarrassing', throwing out the first two versions. In August 2006, Gunns lodged a third version but was again rebuffed. Bongiorno said the new writ was too broad and argued 'too much is being sought to be alleged against too many defendants'.[9] Outside the court Bob Brown was incredulous, declaring, 'it's Gunns that should be in the dock for what it's doing to the forests'.[10] Two months later Gunns' loss was compounded when it was ordered to pay the legal bills of the activists it tried to sue in the first three writs, a cost likely to amount to more than $330 000.

By late 2006, Gunns was starting to look desperate. It was on to its third group of lawyers but still wanted to pursue the writ. The executive chairman of Gunns, John Gay, said the company was tired of having its business damaged and said the company 'will continue to take it through the due processes available to us'.[11] Despite his bravado, however, in December 2006 the company dropped its lawsuit against five of the 20 defendants including Brown, Putt and Doctors for Forests. It also indicated it would no longer pursue its claim that there had been a conspiracy to injure the company by unlawful means as well as several other claims. It later reached an out-of-court settlement with a further defendant.

In April 2007 Gunns went back to court to plead its case for a fourth writ, this time presenting a narrower argument focusing on the company's claim of illegal interference with its trade. Bongiorno allowed Gunns to proceed with the new action, which will be heard in 2008. Although the conservation movement has won the court battles so far, there is wariness about the new writ and fear that those with lower public profiles could end up easy prey for Gunns. Putt voiced these concerns: 'In dropping the claims against Bob Brown and myself this corporate bully has removed defendants with a high profile and left the little people to fight on'.[12] Gunns seems determined to draw the fight out. It is still far from certain who will win.

Gunns' pulpmill

Audacious as it was, the legal action against the forest activists was not the most controversial use Gunns made of its new, corporate muscle. That came in its pursuit of a new pulpmill it wanted to build in the north of the state.

In 2003 Gunns began indicating it was interested in building a pulpmill in northern Tasmania, eventually identifying a site on the Tamar River, not far from Launceston. After the demise of the Wesley Vale pulpmill in 1989, the Tasmanian government had a second major attempt at establishing an export pulpmill in 1997 when it unsuccessfully tried to interest a Taiwanese company in building a pulpmill plant. Six years later Gunns began a third attempt. No doubt mindful of the Wesley Vale mill experience, at first Gunns promised the mill would not use a chlorine bleaching process. It also said the mill would only use plantation-grown timber and would be the 'world's greenest pulpmill'. But the company soon changed its tune. The mill evolved into one that would be chlorine based and would be fed, at least initially, by timber from native forests. The new pulpmill would be the largest private sector investment ever made in Tasmania. It would dwarf earlier pulpmills consuming seven times the timber that the newsprint mill near New Norfolk consumes, and would have enormous economic, social and environmental impacts. It would be one of the largest pulpmills in the world and would have many environmental impacts, but the three that attention became most focused on were its impact on forests, air quality and water quality.

From the environment movement's point of view, the pulpmill's impact on the state's native forests was the greatest concern. The $2 billion mill was touted as adding value to the state's crude woodchip exports, so it was assumed that Gunns would no longer export woodchips once the mill was built. Instead it was thought the company would use its woodchip timber to feed the mill. In 2003 the state's resources minister, Paul Lennon, even said the mill would replace between half and all of the state's woodchip exports. Gunns even argued in advertisements that 'the project is simply based on diverting woodchips that are currently being exported to the pulp mill'.[13] But in August 2006 a regional manager for Gunns told a meeting of logging contractors that long-term woodchip exports were vital to the company's ability to finance the mill. He said that at least until 2025 Gunns would continue to export around 3 million tonnes of woodchips annually, in addition to using up to another 4 million tonnes each year in the mill.[14] The greatest volume of woodchips ever exported by Gunns was 5.2 million tonnes in 2003–04 but the manager presented figures showing that 5.5 to 7 million tonnes of pulpwood would be cut by the company each year, for both its woodchip exports and the pulpmill, once the mill was established.[15] These volumes were confirmed by Gunns' Integrated Impact Statement for the pulpmill which said that the timber harvested by the company from native forest and plantations would reach 6.8 million tonnes by 2017. In many peoples' minds this made the company's claim that 'the pulp mill will not intensify forest operations' questionable, to say the least.

Another early impression given by Gunns was that the mill would be fed by plantation timber and would therefore have no impact on native forests. It became clear, however, that at least for its first decade of operations the mill planned to take up to 80 per cent of its timber from native forests. Forward plans for the mill suggested this would drop to 20 per cent within ten years but this was dependent on plantation growth rates which have generally been below Gunns' expectations. Spokesperson for TWS, Geoff Law, argued the new logging levels would lock in destruction of about 2000 square kilometres of native forest throughout the state. 'This is probably worse than our worst fears because of the sharp rise in destruction of native forests in the first ten years,' Law lamented.[16] A forest scientist from the Commonwealth Scientific and Industrial Research Organisation (CSIRO), Chris Beadle, confirmed Law's fears. Speaking in a private capacity, Beadle explained that there would not be sufficient plantation timber to satisfy all of the mill's timber long-term timber demand. He said there would be a potential shortfall of about 600 000 tonnes by 2020.[17] Gunns' desire to site the mill close to native forest that could feed both the mill and its woodchip operations was probably a major motivation behind its choice of the Tamar River site rather than a less controversial site it considered near Burnie.

As well as placing significant extra pressure on the state's forests the pulpmill would emit air pollutants in large volumes. Criticism of the mill's air pollution centred on its release of particulates, emissions of nitrous oxide and foul odours. The Australian Medical Association was damning of the mill's fine particulate pollution and refused to back the project. A spokesperson for the organisation, Andrew Jackson, said: 'We know that people have been dying over the last two decades as a result of particulate pollution. If we add a further point source into the Tamar Valley, it makes sense that the problem is only going to get bigger.'[18]

In November 2006 it was revealed that a CSIRO review was critical of the mill's air pollution. The CSIRO warned that the mill's releases of nitrogen oxide would exceed official limits, a finding later confirmed by the state government.[19] An expert who reviewed the mill for the state government, Warwick Raverty, even claimed the emissions would breach the Stockholm Convention on Persistent Organic Pollutants.[20] And a report prepared by engineering consultants Beca AMEC for the state government body that initially assessed the mill, the Resource Planning and Development Commission (RPDC), said it was a 'critical deficiency' that the mill proposal did not have any special features that would minimise fugitive emission odours. The consultants also observed that such odours 'have proven to be a significant source of nuisance odours in modern kraft mills'.[21] Experts even told the RPDC 'there is no such thing as an odour-free pulp mill'.[22] As if to confirm the concern about the mill's air pollution, in 2004 the Tasmanian government passed legislation that made all future eucalypt pulpmills exempt from the state's pollution controls.[23]

Another pernicious type of air pollution the mill would be responsible for is greenhouse gas emissions. The enormous amount of timber consumed by the mill would release large amounts of greenhouse gas pollution through its logging of forest and its burning of timber to generate power. An analysis undertaken by Margaret Blakers, from the Green Institute, estimated total emissions of at least 10 million tonnes of carbon dioxide per year: equivalent to two per cent of all of Australia's greenhouse gas emissions.[24]

The third major area of environmental concern was the mill's waste water. The use of chlorine bleaching in the mill would result in the discharge of up to 40 billion litres of effluent containing toxic organochlorine into Bass Strait each year. The mill would consume 72 megalitres of water each day: an amount of water greater than the combined total water use of Launceston and the nearby West Tamar, George Town and Meander Valley areas.[25] Concern grew about the likely impact the dangerous dioxin releases would have on sea life near the mill. Andrew Wadsley, an Associate Professor at Curtin University, warned that nearby seal colonies could suffer birth defects and local shark colonies could be endangered.[26] The Tasmanian Fishing Industry Council said local fish populations could become contaminated and argued the mill would be disastrous for the state's fishing industry.[27] It emerged that even the office of the federal environment minister had concerns. His office said the mill posed a 'credible risk of significant adverse impacts' on threatened marine species.[28]

As happened with the proposed Wesley Vale mill, concern about the environmental impact of Gunns' pulpmill turned many conservative northern businesspeople against it. The mill ended up becoming something of a flashpoint between big business and small business in Tasmania. Much of the tourism, fishing and wine industries came out solidly against it. A report commissioned by the Tasmanian Roundtable for Sustainable Industries group even concluded that the mill would cost 175 jobs in the local fishing industry and 1044 jobs in the tourism industry, costing it $1.1 billion in lost business in its first 30 years of operation.[29] The Tourism Industry Council told Gunns it should stop making dismissive comments and start listening to the legitimate concerns that tourism operators had about the mill.

However it was viewed, there was little doubt Gunns' proposed pulpmill posed a huge environmental threat to Tasmania. It would take industrial-scale forestry in the state to a whole new level.

The pulpmill approval process

In 2004 Gunns' pulpmill began to look like a serious proposition. In June that year, the company initiated a six-month feasibility study of the project, with positive results. In November, Gunns asked the state government to make the mill a Project of State Significance. This status would subject it to assessment by the state RPDC. In February 2005, however, the first piece of intrigue entered the debate about the mill's future. The executive chairman of Gunns, John Gay, announced the mill would not be chlorine-free—as originally promised—but instead would be elemental chlorine-free, meaning the use of chlorine dioxide in the bleaching sequence. Suddenly the environmental credentials of the mill were looking more dubious. Despite this development, Premier Paul Lennon said he would abide by the decision of the RPDC. In June, further intrigue set in when the head of the RPDC's assessment panel for the mill, Julian Green, complained that a taskforce established by the state government to promote the mill was compromising the RPDC's assessment. By the end of 2005, the RPDC's assessment process, while looking fraught, was still on track.

Throughout 2006 the mill's assessment went from being fraught to vexed but was still moving forward. In March that year state elections were held. A few weeks before polling day Gay suggested that if the outcome was a hung parliament—with the Greens holding the balance of power—Gunns might consider taking the mill overseas. Then in November the mill's environmental credibility slipped further when the CSIRO disputed Gunns' air emission forecasts.

Until the end of 2006 the mill was controversial but was still being subjected to a fair and reasonably impartial assessment process. All that spectacularly fell apart in 2007. At the start of the year, two of the RPDC's assessment panel stepped down, including Julian Green. In his resignation letter Green said his position had been made untenable by the ongoing activities of the state government's pulpmill taskforce. Peg Putt, leader of the Greens in the Tasmanian parliament, said 'heads should roll' over Green's resignation and the taskforce should be disbanded.[30]

At much the same time, Gunns started to express frustration over the time the RPDC process was taking. Gay again suggested the company might take the mill elsewhere, this time flagging a possible pull-out should the RPDC process not be finished by the end of June. In early January Gay said: 'Someone's got to make a decision here; if it goes on any longer than six months the whole thing is in jeopardy'.[31] Lennon echoed Gay's sentiments: 'I am concerned about the time it is taking ... I would have hoped we could finish our consideration by the end of this financial year.'[32] Two weeks later Lennon began distancing himself, somewhat, from the RPDC by refusing to promise that he would necessarily abide by its decision. He said his government would keep 'the full range of statutory powers' available to it. In response TWS said: 'this is a spectacular backflip by the premier that should not be accepted'.[33]

Despite Green's resignation, Lennon continued the RPDC's assessment process, appointing retired state Supreme Court judge Christopher Wright to replace him. In late February, Wright said he was 'very optimistic' the mill could be assessed by late November. Gunns accepted his time frame, apparently backing down on its earlier insistence on a June deadline. Wright also pointedly said that 'all or most' of the RPDC's delays to date 'have resulted from Gunns' failure or inability to comply with their own prognostications or the panel's requirements'.[34]

Wright seemed to be trying to be impartial and to steady the RPDC process. Despite this, Lennon met with Wright to see if the process could be sped up. Wright told the premier it was impossible to give a firm date for the panel's decision but would look at where its time frame could be shortened.[35] Nevertheless, he thought Lennon was giving him an 'ultimatum'.[36] He also later told the media he drafted a letter of resignation which may have forced the state government to go along with his time frame. He said:

> I don't know if [Paul Lennon] was doing it off his own back or bat or acting as a messenger boy for Gunns but there was never any doubt in my mind that he was a very enthusiastic supporter of the whole process and that he was anxious throughout that a process that was acceptable to Gunns should be followed.[37]

By early March 2007 it looked as though Wright had succeeded in steadying the RPDC ship but then—out of the blue—Gunns stunned everyone by withdrawing from the process. Gay said the move was prompted by his company's poor communication with the RPDC and its uncertain assessment time frame. He said, 'we cannot continually run a process which has not got any timelines'.[38]

Gay seemed to want Tasmanians to think the withdrawal was entirely related to the RPDC's processes. In June, however, the Greens released documents revealed in a freedom-of-information request that showed a few days before Gunns' withdrawal Wright had intended sending the company a letter saying its documentation to the RPDC remained inadequate and suffered from 'critical non-compliance'. The documents also revealed that the head of Lennon's Department of Premier and Cabinet requested that the letter not be sent, insisting she would take the issue up with the company instead.[39] This new information gave the impression that Gunns' withdrawal may have been prompted by both a wariness of the assessment time frame and of getting an adverse decision from the RPDC.

Paul Lennon is a premier who makes no secret of his support for the forestry industry and he was not about to see the mill disappear once Gunns withdrew from the RPDC process. He spoke with Gay both before, and after, the company's withdrawal then announced he would introduce new fast-track approval legislation for the mill. The special treatment legislation would introduce a new process that would involve his government commissioning two companies to assess the mill's environmental and economic impacts. If they were supportive, he would ask parliament to approve the mill by the end of August. The Greens were aghast. Putt said: 'The Greens are dismayed that cabinet has decided to make a different rule for Gunns than for everyone else in Tasmania'.[40] Lennon kept quiet about Gunns' failure to provide the RPDC with adequate information, and parliament compliantly passed his special legislation without this knowledge (though only after a member of Lennon's Labor Party voted against it and was sacked from the parliamentary party). The legislation closed the ability to appeal the mill's approval under state legislation. A few days after its passage, an independent member of the parliament's upper house claimed a lawyer acting for Gunns told him he had 'been involved in the preparation of the Bill'.[41]

Lennon looked as though he had thoroughly stitched together a new way of getting Gunns' pulpmill approved. But it still had two potential hurdles. One was public support. By May 2007 the mill was starting to look less popular than it once had: a survey taken in the north-east of the state indicated 45 per cent were opposed to it and only 36 per cent were in favour.[42] In June, 11 000 turned out for a TWS rally in Launceston against the mill, one of the largest post-Franklin rallies seen in the state. The other hurdle was federal government approval. Federal assent was needed because the Commonwealth has responsibility for marine waters, endangered species and migratory birds, all of which would be impacted by the pulpmill. After Gunns' withdrawal from the RPDC, the federal government set up a new low-key assessment process whereby the public could send comments to the company which would pass them on, along with its responses, to Canberra. However, in a daring (and expensive) move, TWS announced in May that it would challenge the federal government's new assessment process in the federal court, arguing that it was too limited and lacked rigour. TWS was joined in the action by a new state lobby group, Investors for the Future of Tasmania. The case was heard in July 2007.

In early July 2007 the results of the reports Lennon commissioned for his new approval process were made public. The report on the mill's environmental impacts was compiled by a Finnish engineering firm, Sweco Pic, a business associated with two other companies that would be involved in the mill's construction.[43] Despite finding the mill would not comply with a number of its environmental guidelines—including those relating to nitrous oxide emissions and water effluent—Sweco Pic gave its endorsement.[44] Crucially, Sweco Pic only assessed the mill against half the guidelines that the RPDC was going to use and did not review the mill's testing procedures for toxic pollutants.[45] It also assessed the mill against generalised United Nations Environment Program pulpmill guidelines for an unpopulated area instead of using the RPDC's specific guidelines for the Tamar mill.[46]

The economic benefits report—compiled by ITS Global, which had undertaken publicity work for the logging industry in New Guinea[47]—found the mill had high net social and economic benefits. Incredibly, however, the report did not consider the economic costs of the mill.

With these endorsements in place, Lennon felt he had all the justification he needed for the mill and set about drafting the final legislation needed to get it approved. He also began spending \$300 000 of taxpayers' money to promote the mill through government advertising. In late August his legislation was passed by the Tasmanian parliament. Members of parliament were given little time to review the 1100 pages of state conditions that would apply to the mill and were not able to amend its legislation. In a move that smacked of undemocratic process, the state government allowed Gunns to review the mill's licence conditions before it allowed MPs

Tasmania's first hydro-electric power station at Duck Reach, on the South Esk River. It was established in 1895, just two years after the world's first hydro-electric scheme started at Niagara Falls. (*Archives Office of Tasmania* PH30/7635)

Some of the copper smelters at the Mt Lyell mine. Between 1896 and 1922, they released large quantities of sulphur which formed an acid that killed off much of the surrounding vegetation. (*Archives Office of Tasmania* AA193/2387)

The Opening of Mt Field National Park in October 1917. Along with Freycinet National Park, this was the first reserve created in Tasmania that would go on to become a national park. (*Archives Office of Tasmania* NS869/1/298)

Pioneer of the Cradle Mountain area, Gustav Weindorfer, who was instrumental in getting Cradle Mountain–Lake St Clair National Park proclaimed in 1922. (*F. Smithies Collection, Archives Office of Tasmania* NS573/4/12/1/202)

Albert Ogilvie, the Depression-era premier of Tasmania from 1934 to 1939. He gave the first major push to resource development in the state through a triumvirate of hydro, pulpmill and tourist road development. (*Archives Office of Tasmania* PH30/1/9977)

Electricity gets god-like powers in a 1934 Tasmanian government advertisement. (*The Mercury*)

The establishment of the Burnie pulpmill in 1937 marked the beginning of industrial-strength forestry in Tasmania. (*Archives Office of Tasmania AA193/459 no. 1*)

Jessie Luckman was a leading figure in Tasmania's first major wilderness campaign, the 1940s battle to stop the excision of the Florentine Valley forests from Mt Field National Park. (*The Mercury*)

A 1972 Hobart rally against the Lake Pedder hydro scheme. (*The Mercury*)

Ardent backer of the Hydro Electric Commission, Premier 'Electric' Eric Reece, said the Lake Pedder area contained only 'a few badgers, kangaroos and wallabies, and some wild flowers than can be seen anywhere'. (*The Mercury*)

Lake Pedder before its flooding. Artist Max Angus said about the lake, 'No description, however detailed, could remotely convey the sense of awe and wonder felt by those who saw this magic place'. (*Wilf Elvey, Tasmanian Department of Primary Industries and Water*)

Lake Pedder after its flooding. Former Liberal politician, Edward St John, said of the flooding, 'Our children will undo what we so foolishly have done'. (*The Mercury*)

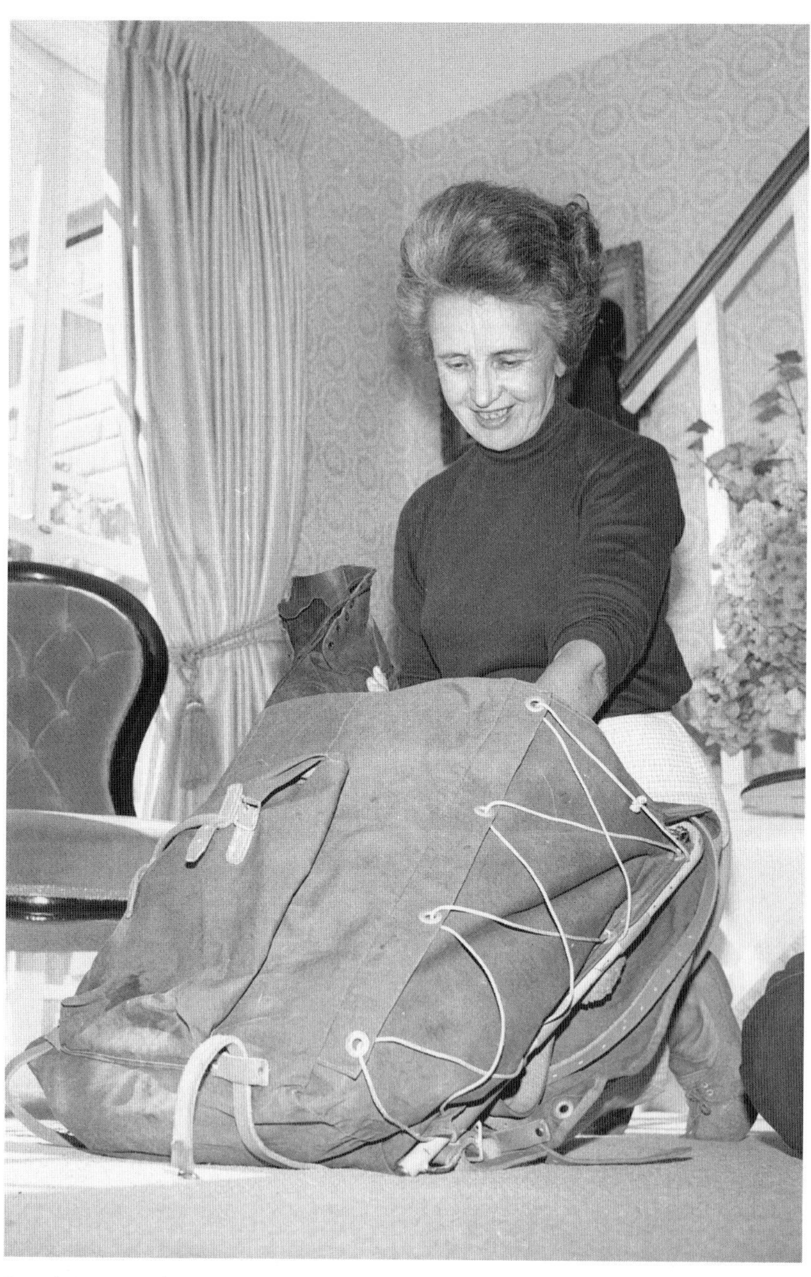

Brenda Hean, a driving force in the Lake Pedder Action Committee, died mysteriously in 1972 while flying to Canberra with Max Price to lobby for Lake Pedder. (*The Mercury*)

In 1970, wilderness photographer and adventurer Olegas Truchanas secured the reservation of a large stand of Huon Pines on the Denison River. He also used his photography to publicise the beauty of Lake Pedder and to campaign for the Pieman River. (*The Mercury*)

Richard Jones was a driving force behind the United Tasmania Group (UTG), the world's first Green party. Like legions of Green party politicians after him, Jones found it hard to get the media interested in the UTG's non-environmental views. (*Patsy Jones*)

Peter Murrell, the first director of the National Parks and Wildlife Service, did not shy away from a fight with the Hydro Electric Commission. (*The Mercury*)

Precipitous Bluff was the subject of four legal challenges mounted by the Tasmanian Conservation Trust between 1972 and 1977. (*Jim England*)

A 1981 Hobart save-the-Franklin rally. (*The Mercury*)

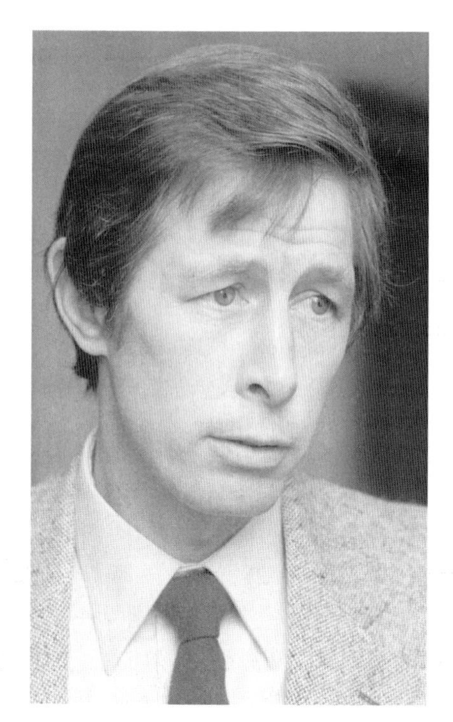

Andrew Lohrey, the best national parks minister Tasmania has had, gave crucial support to the idea of preserving the Franklin River in a Wild Rivers national park. (*The Mercury*)

Premier Doug Lowe's decision to nominate the Franklin River for World Heritage listing in the face of upper house opposition, made federal government intervention to save the river possible. (*The Mercury*)

Bob Brown leaves gaol at the height of the Franklin campaign on the same day he became Australia's first Green member of parliament. As well as leading the fight for the Franklin and establishing the Greens as a national political force, Brown was pivotal in securing a doubling of Tasmania's World Heritage Area in 1989 and saving the Recherche Bay forests in 2006. (*The Mercury*)

Launceston resident Ruth Rowe asks a policeman how she can get arrested at the Franklin blockade. (*The Mercury*)

This is the image that brought the Franklin River into thousands of Australian homes—Rock Island Bend by Peter Dombrovskis. During the 1983 federal election campaign it was used in a national advertisement that asked 'Could you vote for a party that will destroy this?'. (*Liz Dombrovskis, West Wind Press*)

The celebration at the Tasmanian Wilderness Society on 1 July 1983, the day the High Court upheld the validity of federal legislation to stop the Franklin dam. It is still the greatest conservation victory in Australia's history. (*The Mercury*)

Bob Brown being rough handled by forestry workers at the Farmhouse Creek blockade in March 1986. (*The Mercury*)

Future Tasmanian Greens parliamentary leader and senator, Christine Milne, addresses a 1989 Hobart rally against the proposed Wesley Vale pulp mill. (*The Mercury*)

In 1989 businessman Edmund Rouse attempted to bribe a Labor member of parliament so that the Green Independents would not get the balance of power in Tasmania's parliament. (*The Mercury*)

to see them. The government granted twelve changes requested by the company to the mill's operating permits.

The government's fast-track approval process for the mill was never popular in Tasmania. A poll conducted by a commercial polling company early in the same month the mill was approved by state parliament indicated that 64 per cent of respondents disagreed with its approval process while only 25 per cent supported it.[48]

In mid August, the federal court dismissed TWS's case against the federal government's approval process but the organisation appealed the decision to the court's full bench where it also lost, in November. Shortly after the August decision the federal environment minister, Malcolm Turnbull, gave interim approval to the mill. With a federal election imminent the Howard government was concerned it would lose support, as appeared to happen to the Labor Party before the 2004 federal election, if it did not back the forestry industry. (In fact Labor lost support in the north of Tasmania well before its forests package was announced.) There was a major backlash against the announcement of Turnbull's interim approval and a high-profile campaign was mounted against him in his Sydney electorate of Wentworth by prominent businessman Geoffrey Cousins. By late August 2007 the mill was daily front-page news in all the nation's newspapers. Tasmania's wilderness battles were back in the national consciousness in a way they had not been since the Franklin campaign.

Turnbull was under a lot of personal pressure. Keen to find political cover, in August Turnbull asked the federal government's Chief Scientist to investigate the mill's environmental impact and advise on the federal government's response. After receiving this advice, in early October Turnbull announced his decision. To the dismay of the conservation movement, he said the mill could go ahead albeit with 24 new environmental conditions recommended by the Chief Scientist that included new modelling of the mill's ocean impact and oversight of its design by a panel of experts.

However, the new conditions also allowed the amount of chlorate pollution that could be released to be nearly doubled. Peg Putt lamented: 'Mr Turnbull's decision has shone a spotlight on the gross deficiencies of Tasmania's fast-track approval and the restricted nature of the Commonwealth approval.'[49] Geoff Law said: 'The impact of this mill consuming four million tonnes per annum of logs has not been addressed by either state or federal governments.'[50]

Soon after Turnbull's announcement a national election was called. Conservation movement hopes that the federal Labor Party would have a different attitude to Gunns' pulpmill than that of the Howard government were quickly dashed. Despite his background in the conservation movement, Labor's shadow environment minister Peter Garrett said his party supported the decision. Geoffrey Cousins called Garrett 'a shadow minister who doesn't cast a shadow'. Bob Brown said: 'Peter went missing and lost his nerve'.[51]

Both Garrett and Turnbull defended their stances by arguing the mill would be 'world class'. Garrett vowed he would support it 'only if it meets the world's best environmental standards'.[52] Turnbull justified his decision by saying '[Gunns] say this is going to be the cleanest, greenest pulpmill in the world: we will hold you to that promise'.[53] An investigation by *The Age*, however, found the mill would be far from world class: the mill's allowable dioxin release would equal the entire allowable release from all of Sweden's bleached pulp and paper industry, which produces seven times the amount of pulp Gunns' mill would. Erik Nystrom, from the Swedish Environmental Protection Agency, told the paper: 'Why they have set their levels at this level I don't know. Any Swedish mill that saw such levels would be alarmed and act immediately.'[54] Canadian pulpmill dioxin release limits are also significantly tougher than the levels Turnbull allowed Gunns.

By the end of 2007 Gunns was saying it intended starting construction early in 2008. As it turned out, that starting time was similar to the one the company would have had had it stayed with the RPDC process (assuming the RPDC approved the mill). Lennon's fast-track legislation did not speed it up at all.

If Gunns was hoping that opposition would subside once the mill was given federal approval, it was disappointed. In November an opinion poll commissioned by TWS from the polling company EMRS found 54 per cent of Tasmanians opposed the mill while only 37 per cent supported it.[55] TWS also staged a large anti-mill rally in Hobart in November that drew 15 000. The ongoing unease significantly lifted the Greens' vote in Tasmania in the federal election. Turnbull, however, managed to retain his seat. The Greens also recorded a strong vote in other parts of the country, helping to bring down the Howard government and allowing the party to share the balance of power in the senate of federal parliament. Bob Brown said he would use the party's new clout to pressure the incoming Labor government to scrap the mill or at least change its operating conditions. The organisation Tasmanians Against the Pulp Mill also said it was investigating launching a class action against the mill in the courts.

If Gunns was hoping that financial and pulp markets would be enthusiastic about the mill, it was disappointed on that front too. After initially rising on Turnbull's decision, the company's share price soon fell after the federal government approval. Crucially, forecasts of an impending slump in global pulp prices also began circulating after the approval. Gunns planned to sell the mill's pulp into the rapidly expanding Asian market but producers in South America and Asia can produce pulp for two-thirds the cost Gunns can. The forecasts suggested pulp prices would begin a steep decline in 2009 largely because of increased South American supply.[56] Given that the mill would more than double the company's debt level, Gunns can ill-afford the risk of a market downturn. TWS was also lobbying Gunns' banker, ANZ, not to finance it.

The pulpmill is an enormous gamble for Gunns, one it cannot afford to lose. Gunns is a much smaller company than Noranda or Taiwan Pulp and Paper, the two companies that had previously expressed interest in building a Tasmanian pulpmill. The cost of the mill is greater than all of the assets Gunns currently holds so if the mill fails commercially, so does Gunns. Unlike high cost northern hemisphere pulp producers, Gunns cannot sell into its local market and is forced to compete in global markets with low cost southern hemisphere producers it will never beat on price.

At the end of 2007, Gunns bought the land for the pulpmill and for a nearby construction village. Despite this, by early 2008 speculation was increasing that it might not proceed with the mill. The company appeared to be in no hurry to start construction despite the impatience it showed throughout 2007. The mill continued to be unpopular resulting in a low approval rating for Premier Paul Lennon and speculation that he might step down by the end of 2008. In case the mill does still proceed, two new defences were being prepared by the environment movement. One was a legal challenge by the Lawyers for Forests organisation to the final federal approval for the mill given by Turnbull. The group is arguing he did not have sufficient information to make an informed decision. The second defence was in preparation for a possible blockade of the mill's construction. Both TWS and the Tasmanians Against A Pulp Mill group said they had thousands of people ready to join such a direct action. Vica Bayley, from TWS, said protest against the mill would be ongoing. He warned: 'The thing that Paul Lennon and John Gay and Peter Garrett need to realise is that with this pulp mill there will never be an end game.'[57] He also said, 'People must remember that the Franklin dam and [the] Wesley Vale [pulpmill] both had construction commence before they were stopped'.[58]

The future for Tasmania's forests

By 2008 the fight over Tasmania's forests was more than two decades old and showed no sign of abating. Over those two decades the

Tasmanian environment movement has had a number of major wins, including the 1989 doubling of the World Heritage Area, the 1989 creation of the Douglas–Apsley National Park, the 1997 creation of the Savage River National Park, the late 1990s reservation of several hundred thousand hectares of forest as a result of the Recommended Areas for Protection and Regional Forest Agreement processes, and the 2005 gains of the Lennon–Howard forest agreement. These wins have protected the Lemonthyme and Farmhouse Creek forests, part of the Beech Creek and Counsel River forests, many of the Tarkine rainforests and some of the tall trees of the Styx Valley. The movement has also stopped the Wesley Vale and Huon Forest Products mills, has secured an end date for the conversion of native forests to plantations, has stopped the use of 1080 poison on public land and has been responsible for considerable tightening of the state's forestry practices. All are major gains for which the movement can feel justly proud. But after twenty years of campaigning, Tasmania's forests are still very much under threat. While some forests have been saved, those that are not reserved are more threatened than ever before, especially if Gunns' pulpmill is built.

The continuing destruction of Tasmania's forests is particularly lamentable in light of the fact that forest management practices in some other states have left Tasmania behind. When defending its forestry practices, Tasmania could once point to high levels of native forest logging in Queensland, New South Wales, Victoria and Western Australia (the other states and territories do not have significant forestry industries). But half of those states have now halted native forest logging. Queensland, which does not have any woodchipping industry, signalled an end to its native forest logging in the 1998 state election when Premier Peter Beattie pledged to move the state's logging industry into plantation timber. Western Australia has done the same thing; its transition out of native forest logging began in 1999 when Premier Charles Court signalled an end to most of its native trees forestry. The process was hastened by his successor, Premier Geoff Gallup, who won a state election in 2001 on the back of a pledge to immediately stop all native forest logging.[59] Stuck in the last century, Tasmania is going against the trend by continuing to log native forest.

Luckily, many of the companies that buy Tasmania's woodchips are ahead of the state forestry industry. In 2004 both the Mitsubishi Corporation and Ricoh announced that they would no longer buy woodchips that came from Tasmania's old-growth forests. The move away from native forest timber is also apparent in woodchip trends. A growing preference for plantation-sourced hardwood woodchips was the driving force behind a 26 per cent fall in Gunns' woodchip exports from Tasmania between 2004 and 2006. Many observers argue this was a major motivation for the company wanting to build a pulpmill: it needed a market for old-growth woodchips that overseas buyers increasingly did not want.

Only 30 per cent of the forests that existed when Europeans first settled in Tasmania are protected. The other 70 per cent has either gone, or may soon go. As with so many of Tasmania's resources, there is an unwritten law in the state that forests exist for industry use unless there are overwhelming reasons otherwise. This law has huge defences around it made up of Forestry Tasmania's legislated control of the state forestry industry; an unholy alliance between the state government, Gunns and forestry unions; national Regional Forest Agreement laws that exempt environmental protection of forests; weak independent regulation of forestry in Tasmania; and a legislated requirement that at least 300 000 cubic metres of sawlogs and veneer logs will be cut each year in the state. These defences give immense power to the industry and the conservation movement has done well to defy it to the extent it has.

The din of the chainsaws continues and Tasmania's tall trees keep coming down like Gulliver yielding to the might of a thousand Lilliputians. Eagles, parrots, bettongs, possums and bandicoots keep fleeing the advancing clearfelled coupes and the forested

Cartoon by Ron Tandberg that appeared in *The Age* and *The Sydney Morning Herald* in 2007 commenting on Gunns' proposed pulpmill.

boundaries of the state's wilderness areas keep shrinking. Tasmania's forests are still some of the most inspiring in the world but the threats to them remain great. Tasmanian wilderness photographer Olegas Truchanas once said: 'If we can accept the role of steward, and depart from the role of conquerer; if we can accept the view that man and nature are inseparable parts of the unified world—then Tasmania can become a beacon in a dull, uniform and largely artificial world'.[60] Tasmania has yet to become a steward. It has yet to find the humility to stop itself from conquering its forests.

Far from being humble, in March 2008 John Gay said Gunns was confident of high future pulp prices and would start work on the pulpmill in July. This set the stage for either a humiliating backdown by Gunns or the biggest blockade Tasmania has seen since the Franklin campaign.

MINING

8

Mount Lyell to Benders Quarry

Mining has had a lower profile as a destructive force on Tasmania's wilderness areas than hydro development or forestry. Few would probably be aware that mining has had any impact at all on the state's wild areas. In fact it has had enormous impact. Large-scale mining began in Tasmania earlier than industrial forestry or hydro development so much of the industry's impact occurred before there was much environmental awareness. Nevertheless, mining has had a devastating effect on Tasmania's pristine natural areas and has left a sorry legacy in many once-wild parts of the state.

Throughout most of the nineteenth century, western Tasmania largely escaped any development. It was considered an inhospitable, mysterious place often referred to as Transylvania. The only major developments that penetrated its curtain were a short-lived penal colony on Sarah Island, in Macquarie Harbour (halfway up the west coast), that operated from 1822 to 1833, and some settlement and wool-growing activity undertaken by the Van Diemen's Land Company in the far north-west that began in 1826.

The event that forever changed the splendid isolation of western Tasmania was the discovery of a rich tin deposit by James 'Philosopher' Smith. Smith had taken part in the Victorian goldrush. In 1853 he moved to northern Tasmania, where he spent most of the following two decades searching its highlands for an elusive 'El Dorado'. In 1871 his perseverance paid off when he discovered a major tin deposit at Mount Bischoff.[1] His discovery sparked a mining boom in Tasmania and led to the development of one of the largest tin mines in the world as well as the discovery of other major tin deposits throughout the state. It also resulted in an unprecedented wave of prospectors descending on western Tasmania. Suddenly there were shanty towns in areas that had been thick rainforest. Horse tracks began snaking their way along remote ridges that had hitherto known no human visitation. And the area's myriad valleys came alive with miners attempting to repeat Smith's luck. All of the minerals found in the west were discovered in a long set of highly mineralised mountains running down the western side of the state that came to be known as the 'Mount Read volcanic belt'. This belt became holy ground to the state's mining industry, whose sanctity it fiercely defended against later conservation threats, especially possible inclusion in national parks.

The Mount Lyell copper mine

Ten years after Smith's strike gold was discovered on the King River, which flows into Macquarie Harbour along with the Gordon River. Two years after the discovery, prospectors William and Michael McDonough and Steven Karslon climbed onto a ridge above the King and came across a large outcrop of iron at a place called Mount Lyell, near present-day Queenstown.[2] The miners thought gold must surely lie below the deposit but later prospecting revealed a huge copper vein to be situated there instead. In 1892 the Tasmanian parliament passed legislation allowing the deposit to be developed and soon afterwards the Mount Lyell Mining and Railway Company began exploiting the lode. The mine went on to become one of the largest copper mines in the world.

A key part of developing the ore body was smelting it into a relatively pure form. A US metallurgist, Edward Dyer Peters, advised the company to use a smelting process that used a 'pyritic' system and in 1896 the fire was lit in the furnace of the mine's first such smelter.[3] By the turn of the century, ten more smelters using the same process were operating near the mine. Although hugely successful in technical and economic terms, a significant pernicious by-product of the smelting was its large releases of sulphur. After mixing with moisture-laden air the sulphur emissions formed sulphuric acid. The smelter chimney stacks were therefore responsible for thick clouds of dilute sulphuric acid settling around the mine and killing everything they fell on—they were clouds of death. When all the smelters were operating they churned out a cocktail of 130 000 tons of the acid each year.[4] The emissions wiped out vast areas of vegetation, stark evidence of which can be seen in the eerie, denuded moonscape around Queenstown to this day.

It was soon realised how deadly the pollution was and all the miners who lived near the smelters were relocated to the embryonic settlement of Queenstown.[5] Historian Geoffrey Blainey wrote in his definitive history of the Mount Lyell mine, *The Peaks of Lyell*, that: 'Sulphur was the curse of Mount Lyell. When the big company smelted its pyrite in ten or eleven large furnaces Queenstown found its climate changing. In still weather sulphur from the smelters thickened fogs into pea-soupers, choked Queenstown, and blanketed the valley.'[6]

In the 1920s the Mount Lyell mine fell on hard economic times. It eventually traded its way back into profit after halting the pyritic smelting process in favour of a more efficient 'floatation' smelting process that it began using in 1922. The new smelting process was both a blessing and a blight. It no longer released sulphuric acid clouds and both people and the environment could breathe more easily as a result. However, the mine now produced huge quantities of waste tailings, which were dumped straight into the Queen River, a tributary of the King. The tailings discharge into the Queen River ramped up from 100 000 tonnes in 1922 to 1 500 000 tonnes in 1935. In the 1960s and 1970s they reached a massive 2 000 000 tonnes a year.[7] The local air was spared but the nearby rivers were poisoned. No consideration whatsoever was given to the environmental impact of the mine waste that poured into the river. The mine was all-important, the waterways were not. Even when the Tasmanian government passed its *Environment Protection Act* in 1973 the mine was given an exemption and the dumping of the tailings continued.[8]

The result was that by 1994, when a tailings dam at the mine was finally built, an estimated 95 million tonnes of sulphidic waste had been poured down the Queen and King rivers along with 1.4 million tonnes of smelter slag and 10 million tonnes of topsoil.[9] The rivers became sterile. No aquatic life could survive in them and very little vegetation could grow along their banks. They are the most devastated waterways in Australia and are an abiding disgrace of the Tasmanian mining industry. Old photos of the King River, taken before the mine began, show a majestic river whose beauty rivalled that of other western Tasmanian rivers like the Gordon, the Pieman and the Franklin. But today the King is an environmental graveyard, an epitaph to the folly of men determined to win wealth from the earth regardless of environmental cost. Probably the most graphic legacy of the mine discharge is a huge 250-hectare delta that has developed at the mouth of the King River where it enters Macquarie Harbour. After rushing down the river, most of the mine tailings stopped when they reached the harbour then accumulated to form the delta. It contains about 85 per cent of all the mine waste discharged over nearly a century of mining.[10]

In the late 1970s and early 1990s low copper prices again threatened the viability of the mine and in 1993 work stopped for a short time. In 1994 a change in ownership was the catalyst for the state government to finally insist on the construction of a tailings dam, and for the first serious examination of what could be done to rehabilitate the King. Studies of the river identified the main environmental threat as ongoing discharges from the mine site of large volumes of water that were highly acidic and laden with copper. Water entering the river was flowing over sulphide-bearing tailings rocks, which was oxidising to form acid. Most of the drainage came from disused underground mine workings. Chemical and toxicological testing indicated nearly all of the acid drainage had to be neutralised before the river could meet basic environmental standards. But detailed examination of the costs of extracting the copper found it was cheapest to extract it without neutralising the water. This would improve the Macquarie Harbour ecosystem (where there is significant fish-farming activity) without yielding much, if any, improvement in the quality of the King's waters. Simply removing the copper would have a twenty-year operating cost of $16 million compared to a full neutralisation cost ten times greater over the same period.

The state and federal governments, along with the new mine owners, ultimately settled on simply taking out the copper, a process that will begin in about 2009. They put major rehabilitation of the King and Queen rivers into the too-hard basket. Other initiatives, such as redirecting clean water that came out of the mine site and some revegetation of the King River–Macquarie Harbour delta, have also been undertaken. But the really serious programs of fully neutralising the river water, and possibly dredging some of the tailings, have been dropped because they are deemed too costly. Fixing a century of abject mining abuse was never going to be cheap but it is very sad that the comprehensive rehabilitation of the river has been dismissed as too difficult and will have to be revisited by future generations.

The huge cost of fixing the environmental legacy of the Mount Lyell mine was exacerbated by the fact that, like most mines in Tasmania, it paid no royalties to the state government until the 1970s. Even when it did pay royalties they were lower than those levied in any other state or territory. When an attempt was made by Premier Eric Reece in the mid 1970s to levy royalties more in line with the national average, the Mount Lyell mine took its plight to the media. The Hobart *Mercury* ran the headline 'Lyell Mine

Threatened—Plea to Cut Royalties', which forced Reece to back down fearing the industry could suffer.[11] When the mine fell on hard times in 1978 the state and federal governments even offered to underwrite its losses.[12] So the Mount Lyell Mining and Railway Company has made little direct contribution to the government purse. It was a liability to both the environment and the state's public finances.

Today a major scenic railway that once transported ore from the mine takes tourists along much of the lower reaches of the King River. Instead of an untamed river its passengers witness a toxic requiem to the disregard of miners and businessmen past.

Tin mining at Melaleuca

Apart from the Mount Lyell mine, another legacy of James Smith's 1871 tin discovery was the discovery of tin twenty years later at Cox Bight, in the far south-west corner of Tasmania. Cox Bight is a stunning coastal wilderness area surrounded by large, remote mountains. Although carried on at a modest scale, the start of tin mining in the area by the Freney Prospecting Company put a pinprick in the middle of the state's most significant wilderness. Eventually the mine area was slightly expanded after farmer and sawmiller Charles King joined the syndicate that worked it in 1933.[13]

The following year, the last major discovery of tin in Tasmania was made at Melaleuca, 8 kilometres north-west of Cox Bight, situated on the edge of the majestic Bathurst Harbour. The Melaleuca deposit was initially developed by the New Harbour Tin Company but in 1941 was taken over by King, who was joined by his son, Deny, four years later. The Kings continued to work the deposit for the rest of their lives. Like the Cox Bight mine, the Melaleuca mine never grew to be more than a small operation carried on by a maximum of three or four people. But, like Cox

Bight, it pierced the otherwise unpenetrated cloak of remoteness that protected the south-west.

Eventually mining at Cox Bight ceased but tin mining at Melaleuca continues to this day. Charles King died in 1955 although Deny kept mining at Melaleuca until his death in 1991 and in 1974 was joined by fellow miner, Peter Willson, who is still mining today.[14] Deny eventually made Melaleuca a small settlement complete with several houses, bushwalker huts and an airstrip, which he completed in 1957. He went on to become a much-loved character of the south-west wilderness, who did a lot to popularise and preserve the area. He was instrumental in securing reservation of the nearby Port Davey foreshore in 1961 and in the late 1980s gave invaluable assistance to saving the threatened population of rare orange-bellied parrots that nested in the area.[15] But however much King was admired and respected, it is difficult to escape the fact that tin mining scarred the area and had a significant environmental impact. Although topsoil was put back over the mine workings they still impair the Melaleuca landscape. King was sympathetic to some form of rehabilitation but did not take kindly to directives given to him by the state environment department in the 1980s about how the workings should best be rehabilitated.[16]

Worst of all, the area's tin mines ensured that it was never incorporated into the Southwest National Park nor included in the Tasmanian Wilderness World Heritage Area. Over the years the Parks and Wildlife Service has made several approaches to the State Mines Department seeking to at least reduce the area of the Melaleuca–Cox Bight mining lease but all approaches have been rebuffed. The mineral exploration lease in the Cox Bight area expired in the mid 1990s. In 1996 an appeal was made to the state environment minister, Peter Hodgman, by the Friends of Melaleuca conservation group, asking for the area to be incorporated into the surrounding World Heritage Area but he said he lacked sufficient cabinet support for the move.[17] As at Mount

Lyell, mining began at Cox Bight and Melaleuca during a period of negligible environmental awareness but, just as the Mount Lyell mine pollution of the King River needs to stop, so too does tin mining in the south-west. The Melaleuca area should be fully rehabilitated and incorporated into the surrounding national park. Some conservationists have even called for the closure of its airstrip. Underscoring the need to properly protect the area was an application for a new Cox Bight mining lease made in late 2007.

The Melaleuca–Cox Bight area was not the only part of the south-west to be affected by mining. On the Jane River, which flows into the Franklin River, gold was discovered in 1934. By the following year 33 men were working the deposit. Small-scale mining continued there for most of the next 50 years despite difficult access.[18]

In 1909 a significant mining operation began at a town called Adamsfield near the Florentine Valley and the middle reaches of the Gordon River. Osmiridium, a corrosion-resistant alloy used in pen nibs, light filaments and the manufacture of industrial jewellery, had been found there. In 1925 and 1926, when prices of the metal were high, up to 2000 men worked the area's soils for the alloy.[19] But the activity was short lived due to exhaustion of the best deposits and a fall in prices. By the beginning of the Second World War virtually all work had ceased, although some mining continued after the war.

Mining in Cradle Mountain–Lake St Clair National Park

Another site of mining activity that might have also led to the rupture of a major wilderness area was the Pelion region in the middle of Cradle Mountain–Lake St Clair National Park. In 1938, wolfram (used in the making of tungsten) was discovered in the area, which sixteen years earlier had been reserved in the national park. (In the 1890s copper was also discovered in the area, before it was made a national park.) The government of the day—headed by Tasmania's pro-forestry, pro-hydro, pro-mining premier Albert Ogilvie—saw little problem in allowing the mining of wolfram to go ahead in the national park. But the Scenery Preservation Board—the authority that managed the park and the act it was created under—was reluctant. The board thought it would set a dangerous precedent and put up a gallant fight against the mine.[20] The board was no match for Ogilvie, however, and the premier succeeded in allowing the mine to go ahead, after his government watered down the *Scenery Preservation Act* under which the national park was created. The amendments made it clear that parts of the act could be suspended, or revoked, to allow mining to take place in a reserve.[21] In the end the act was suspended over 1300 hectares of the national park. Fortunately, the mine was not profitable and closed in 1944 with the excised land eventually returned to the park. Had the mine been more lucrative, however, it would have led to ongoing mining activity in one of Australia's best-known national parks.

Mining in Freycinet National Park

Another national park that has seen mining allowed within its borders is Freycinet National Park. Freycinet, along with Mount Field National Park, was in 1916 the first national park created in Tasmania (see Chapter 9). In 1935 the state government allowed a granite mine to start operating which, in 1941 and 1942, led to the revocation of 6 hectares of the park. The revocations allowed the mine to continue for the next 50 years. In the 1990s the mine wound down before rehabilitation work began and, eventually, the excised land was returned to the national park. Like the Pelion mine, it beggars belief that the mine was ever allowed to be estalished within the park but as has so often been the case in other Tasmanian reserves, national park status was no guarantee of protection.

Iron ore mining at Savage River

Savage River is a remote mining town in the north-west of Tasmania situated about 15 kilometres from Savage River National Park. Surrounded by steep mountain ridges predominantly covered in primeval rainforest, it is in the southern part of the Tarkine wilderness—the largest area of temperate rainforest in Australia. Since the mid 1960s a large iron ore mine has operated in the area. It has had a huge impact on the Tarkine environment, worsened by a long pipeline that was built in the 1960s from the mine, through the surrounding rainforest, to the north coast. At the coastal end of the pipe an ugly processing facility was also constructed. The Mount Lyell, Melaleuca, Pelion and Freycinet mines were all established well before there was much environmental awareness in Tasmania but the Savage River mine was not. It is a relatively new mine whose wilderness impact could have, and should have, been avoided.

Like most of western Tasmania, the Savage River area had been explored for minerals since the late nineteenth century. Little was found then or in the early twentieth century but in 1956 the Mines Department explored the area once more. This time it found paydirt and its favourable assessment resulted in the Rio Tinto company taking out a prospecting licence in the area.[22] In 1961 the company's mining lease was taken over by Roy Hudson, who had previously established a uranium mining company at Mary Kathleen in Queensland. In 1963 Hudson undertook a feasibility study of commercial mining at Savage River with Pickards Mather International, a United States-based company with experience in working low-grade iron ore deposits.

In 1965 Hudson and Pickards Mather International announced that it thought a mine in the area could go ahead. It was envisaged that a concentrating mill would be built at Savage River that would pump ore slurry to a pelletising plant at Port Latta, on the north coast, whose infrastructure would include a bulk carrier loading facility.[23] The government of Premier Eric Reece (who had been a miner in his younger years) embraced the proposal with enthusiasm, offering the mine's proponents extraordinarily generous inducements including a $4 million interest-free twenty-year loan; government construction, or improvement, of roads, powerlines, schools, medical facilities, police offices and other infrastructure at Savage River; and the right for the mine and pelletising plant to discharge waste water into Savage River and the waters around its north coast plant.[24] The Reece government also had no qualms about allowing the mine's 90-kilometre pipe to be built straight through the Tarkine wilderness. Acting as though no lessons had been learned from the Mount Lyell experience it poured huge amounts of money into the project, which had large environmental costs but little net economic value.

Today, when you stand at the western edge of Rocky Cape National Park and look west, you see a long, beautiful stretch of coastline punctuated by a giant, ugly dock. When you drive past the Port Latta processing plant situated at the end of the dock you see a sooty, dank mill reminiscent of the worst excesses of the Industrial Revolution. The Savage River mine has imposed a massive scar on one of the remotest parts of the state. Also, like the Mount Lyell mine, it acidifies nearby river systems. The acidification problem is not as severe as at Mount Lyell, and it is hoped the discharge can be neutralised, but it is saddening that the same mistakes keep being made over and over again.

The Precipitous Bluff fight

Precipitous Bluff is a large, towering dolerite rock ridge that rises out of Tasmania's south-west wilderness, several days' walk from the nearest settlement. Upon reaching its 900-metre summit one is rewarded with an amazing panorama that takes in cliffs that plunge down to a neighbouring lagoon; wild seas and isolated islands off the nearby south coast; scores of surrounding remote, jagged peaks separated by deep valleys; and a sky that stretches forever. In the 1970s the bluff became the subject of a gruelling battle between the Tasmanian Conservation Trust (TCT) and a company determined to extract minerals from the area.

In 1968 Precipitous Bluff was conspicuously excluded from the new Southwest National Park, despite argument to the contrary from the conservation movement. The following year a mineral survey reported that a large band of limestone occupied the lower slopes of the mountain. The Tasmanian environment movement did not view the find as a threat until December 1971 when a Melbourne-based prospecting company—Mineral Holdings—advertised that it was seeking a licence to mine at the bluff.

The idea that mining could be contemplated at such a remote wilderness location caused alarm in the conservation movement. The Australian Conservation Foundation (ACF), the Society for Growing Australian Plants, the South West Committee, the Hobart and Launceston walking clubs and the TCT all formally objected to the granting of the licence. For the ACF and the TCT—which had both been criticised for their lack of interest in the fight to save Lake Pedder—it was a refreshingly assertive stance to take. When the objection was heard in the State Mining Warden's Court in December 1972, Mineral Holdings argued that the TCT and the other objectors had no direct, legal interest in the area since they were not adjoining landowners. This became a crucial point after Reece's state government stopped the National Parks and Wildlife Service, which was an adjoining landowner, from objecting. In a ruling that made state legal history, the Mining Warden found the objectors did have standing and refused the licence. As well as being premier, Reece was minister for mines and had the power to overturn the ruling, but before he could Mineral Holdings lodged an appeal with the state Supreme Court.

The appeal was heard in June 1973. By then the issue had become the most significant campaign the TCT had ever handled.

The President of the TCT, Keith Vallance, also noted that, in contrast to the division within the conservation movement during the Lake Pedder fight, the Precipitous Bluff campaign had 'unified conservation organisations in Tasmania in a way that Lake Pedder failed to do'.[25] The TCT's campaign assumed a high profile and even received support from the parliamentary Liberal Party opposition. The TCT was baffled by Mineral Holdings' determination to pursue the mine. Precipitous Bluff is in a remote part of the state that has no road or shipping infrastructure. The TCT estimated that the cost of limestone taken from the area would be two and a half times that of limestone extracted elsewhere.

Sadly for the TCT, in May 1973 the Supreme Court overturned the Mining Warden's decision. Pat Wessing, a spokesperson from the TCT, called the decision a 'narrow, nineteenth century view'.[26] Despite the setback, however, the Trust decided to battle on and appealed the decision to the full bench of the state Supreme Court (aided by federal government supplied Legal Aid funds). It was two years, however, before the full bench heard the appeal. The delay was used as an excuse by the state government to hold off publishing a new draft management plan for the Southwest National Park that was likely to recommended expanded boundaries. Finally, in May 1975, the court announced its decision: all three judges upheld the earlier decision and Mineral Holdings' costs were awarded against the TCT. The Trust was devastated. It had staked a lot on the case and had lost, for a second time.

Shortly before the decision, the new draft Southwest National Park management plan was released. Its most controversial recommendation was that Precipitous Bluff should be included in the national park but that a large part of the nearby Hartz Mountains National Park should be excised. The excision was to be compensation for the forest resource the Australian Paper Manufacturing (APM) company would lose from the inclusion of Precipitous Bluff, an area where it had 'concession' rights to the pulpwood. The recommendation was controversial because the lack of any forestry roads near the bluff made it unlikely its forests would ever be profitable for APM to log. Also, under the terms of its forest concession it was not clear that the company was entitled to any compensation. APM basically saw the same opportunity that Australian Newsprint Mills had seen in the 1940s when it successfully pressured the state government to excise forests from Mount Field National Park. And, as had by then been the case with Cradle Mountain–Lake St Clair, Freycinet and Lake Pedder national parks, as well as in the 1940s with Hartz Mountains National Park, the Tasmanian government was all too willing to sacrifice the integrity of a national park for corporate gain. National park status meant nothing.

The TCT was delighted with the inclusion of the bluff in the park but had a mixed reaction to the excision from Hartz Mountains National Park. Vallance said: 'I am appalled that this inclusion has resulted in a "trade-off" so that almost half of the Hartz National Park is now to be exploited for forestry purposes'.[27] But others in the TCT were more qualified in their opposition. Some came to see it as a necessary evil and felt there was little alternative. Although not officially supported by the TCT, some of its members actively lobbied the state government in support of the swap. This allowed the government to claim that a major part of the conservation movement—including the TCT, the Hobart Walking Club and the South West Committee—supported the arrangement. Younger, more radical members of the movement were unable to get their voice heard by the government.

Though the acceptance of the swap by some in the TCT was no doubt the result of fatigue after a long and unsuccessful campaign, many in the environment movement saw it as a sell-out. It is still a controversial outcome. After the decision, younger activists set about infiltrating the TCT but met with limited success. Like any broad church, the Tasmanian environment movement includes idealists and pragmatists and both had a valid point of view about the swap. But it taught the movement a lesson about accepting poor political tradeoffs.

By the time of the Supreme Court's full bench decision, the TCT had been campaigning for three years for the right to be able to object to mining licences and could have reasonably been expected to wind up its campaign. However, following a legal opinion from Edward St John (who had been a member of the federal government's inquiry into the flooding of Lake Pedder), and the promise of more support from Legal Aid, the TCT decided to bat on and appealed to the High Court. It took another two years for the High Court to hear the appeal. A three-to-two majority ruled against the TCT in 1977. A disappointed TCT President, Bruce Davis, said: 'once more, the sheer conservatism of the law would seem to fly in the face of social realities'.[28]

The organisation vowed to keep pressuring to change the rules of court standing and, eventually, this occurred. In 1982 the Tasmanian Wilderness Society had little difficulty in securing standing for a High Court case it brought against federal government loan funds provided for the Franklin dam, while in 1994 the TCT established standing in a successful federal court action it brought against the federal government granting of a Tasmanian woodchip licence to timber company Gunns. Although eventually vindicated by the change in court standing laws, the five year campaign left the TCT exhausted and saddled it with a large legal bill above and beyond the costs met by Legal Aid. For the TCT the appeals were more about establishing the right of environment groups to be represented in court than they were about saving the bluff, which came about through the national park boundary change.

Mining and the World Heritage Area expansion, the Salamanca Agreement and the Regional Forest Agreement

In 1989 intense negotiations began between the state Green Independent members of parliament, headed by Bob Brown, and the minority Labor government of Premier Michael Field, over an expansion of Tasmania's World Heritage Area. The result was a doubling of size of the World Heritage Area, mainly to conserve greater areas of old-growth forest (see Chapter 5). And even though the expansion was driven by the need for greater forest protection, the mining industry managed to exert considerable influence over its final shape—Field made sure the Mines Department had a seat at the expansion talks.

Brown had hoped the enlargement would push further west than it ended up doing and that the Tyndall Range, outside the south-west boundary of Cradle Mountain–Lake St Clair National Park, as well as the areas near the west coast below Macquarie Harbour, would be included. But as these two areas are in the mineralised Mount Read volcanic belt that is valuable to the mining industry, the Mines Department advised that they could not be included. Field backed the department. He told Prime Minister Bob Hawke that his government 'firmly believes that these areas have mineral potential and should be available for future development'.[29]

A similar position was taken on the new Douglas–Apsley National Park, which was created at the same time. The mining industry had never shown a commercial interest in the coal deposits under the park but it was keen, nonetheless, to retain access. The industry convinced the Field government to proclaim the park with a proviso that mining could take place underneath it at a future date, thereby ensuring that mining was not locked out.

In 1989 as part of the Salamanca Agreement, negotiations commenced between the environment movement and the state forestry industry over alternatives to logging Tasmania's National Estate forests (see Chapter 5). These discussions were a rare attempt to reach consensus in Tasmania's forests debate, but failed the following year after the Field government began to distance itself from the process. One of the pressures that led to the collapse of the talks was an insistence by the mining industry that it be included in the discussions so that it could make sure no mineralised areas would be reserved. Premier Field acceded to this requirement, giving the industry a veto over any new reserves the process might recommend.

The mining industry was as vigilant about the potential for the 1997 Regional Forests Agreement (RFA) to lock it out of prospective areas. The RFA resulted in about 300 000 hectares of new forest reservation. This was far less than the conservation movement had hoped for. To make matters worse, only 60 000 hectares were protected as national park or state reserve; the rest was reserved as conservation areas, regional reserves or state forest, all categories that allowed mining. National parks are the most secure form of protection a reserve can have in Tasmania; other types of reserve extend lesser protection. Yet again, the mining industry ensured it was not disadvantaged.

Benders Quarry

Even after the expanded boundaries of the World Heritage Area were settled in 1989, mining continued to wield influence on the region. One of the conditions the Tasmanian government placed on the World Heritage expansion was that an existing mine within the expanded boundaries be allowed to continue operating. That mine was Benders Quarry, a limestone mine (like that proposed for Precipitous Bluff) situated near Lune River. The conservation movement initially saw the mine as the price it had to pay for the World Heritage expansion but became increasingly concerned as evidence mounted of the damage the operation was doing to Exit Cave—the longest underground cave system in Australia.

There had been concerted efforts in 1970 to reserve the area around the 20-kilometre-long cave but they only resulted in modest reservation. The 1989 expansion of the World Heritage Area resulted in the first meaningful inclusion of the cave system but the Benders Quarry operation remained a problem. In 1992 The

Wilderness Society began publicising the impact of the mine on the cave and held a number of protests at the site. It also began lobbying the federal environment minister, Ros Kelly, to use her World Heritage management powers to close it. The state government wanted the mine kept open, and after considerable tussling between it and the federal government Kelly acted in late 1992, declaring that the mine would close the following year. The move sparked major protests from the state government and federal Liberal Party but Kelly did not budge. The closure went ahead with later rehabilitation.

Although a lower-profile resource industry in Tasmania than hydro development or forestry, mining has had a long and deleterious impact on the state's wilderness. Its pervasive influence has all too often resulted in the despoliation, and inadequate reservation, of major wild areas. Now that there will be no more dams in Tasmania, surely the time has come when there should be no more wilderness mines. The Tasmanian government must say 'enough is enough' to the mining industry.

NATIONAL PARKS

9

The Scenery Preservation Board

In nearly all of Tasmania's wilderness battles the ultimate aim of the conservation movement has been to preserve threatened areas as national parks. Superficially, it looks as though it has been a successful approach. A large proportion of Tasmania, nearly 1.5 million hectares, or a bit over one-fifth of the state, is reserved in national parks. Because of this, Tasmanians are constantly reminded that more of their state is reserved in national parks than in any other Australian state.

As early as 1968, when the battle over Lake Pedder was raging, the minister for Lands and Works, D.A. Cushion, said: 'from the latest statistics available, Tasmania has a higher percentage of its area reserved as national park and scenery reserve than any other state in Australia'.[1] This was a line countless Tasmanian politicians after him would use. The problem with the argument is that large segments of the state's national parks are in areas no one wants to log, flood, mine or build a tourist lodge on. They do not contain economic resources and they are not representative of all the state's ecosystems. So to understand Tasmania's wilderness battles one has to understand its national parks and the often convoluted history that has led to their creation.

Nineteenth-century reserves

Although wilderness and species protection are common motivations for the creation of national parks today, they were not necessarily the main influences in the nineteenth century when Tasmania's first reserves were created. During the mid to late nineteenth century, when interest in the environment first emerged, a host of influences were in play around the world. The most significant was the Industrial Revolution, which saw the proliferation of factories in often heavily polluted cities that created a yearning amongst city-dwellers for unspoilt places. At the same time there was a railway construction frenzy in most western nations which gave people the means to escape their stifling cities.

In Tasmania in the 1860s the Royal Society began taking an interest in the observation and protection of wildlife. In 1888 a state ornithological society was established, followed by the creation of the Tasmanian Field Naturalists' Club in 1904.[2] At the same time, expressions of concern about the human impact on the environment were also emerging. One such expression was the 1864 publication of *Man and Nature* by George Perkins, which warned of the environmental impact of forestry, agricultural land clearing and industrial pollution.[3] As well as fostering a concern for the environment, in Tasmania these influences helped create a significant tourism industry because the state's vast unspoilt lands were seen as an escape from development. They were spurred along by the 1850s Victorian goldrush and by the state's reputation for having a healthy, 'European' natural environment.[4]

These forces coalesced into pressure on the state government to introduce legislation that would enable the reservation of natural areas as well as protection of wildlife. This resulted in the passing of the ironically named *Wastelands Act* in 1858 and the *Protection of Native Game Act* in 1860. Before 1858 there had been no formal legislative mechanism for the setting aside of land for public purposes, even for city parks.[5] The best part of another 30 years passed before the act resulted in the reservation of any significant areas. In 1885 Russell Falls, in today's Mount Field National Park, as well as a number of lakes on Tasmania's Central Plateau, were reserved under the legislation, largely as a result of recommendations from the Royal Society and a number of government surveyors. By the end of the nineteenth century, other small reserves were created to protect coastal scenery around the Tasman Peninsula, in the state's south-east, and some caves in the north.[6]

In 1903 the *Wastelands Act* was superseded by a new *Crown Lands Act,* the provisions of which were expanded in 1911 to allow for the easier creation of natural reserves.[7] After the start of the twentieth century, other new reserves included a reserve on the alpine Ben Lomond Plateau, in the north-east of the state, as well as a reserve at Rocky Cape, on the north-west coast, proclaimed to conserve some of its unique population of *Banksia serrata.*[8] These two reserves, along with the one at Russell Falls, would go on to form parts of later, larger national parks.

Despite this flurry of reserve creation early in the century, by 1910 only eleven reserves had been created, covering just 10 400 hectares. Their combined size was equal to less than 1 per cent of the area that would be reserved by the start of the twenty-first century.[9] The Tasmanian government was still a long way away from reserving significant areas of the state's wilderness, in large measure because tourism—not protection of the environment—was invariably the main reason for the reservations. The northern caves, for instance, were reserved after lobbying from the Northern Tasmanian Tourist Association.[10]

Creation of the Scenery Preservation Board

By the start of the First World War it was obvious Tasmania needed a better system for both the creation and management of its reserves. More and more tourist and natural science organisations were being formed that had increasingly higher expectations of the state's reserves. There was also mounting evidence of mismanagement of Tasmania's reserves. Hunting was taking place within a fauna reserve on Freycinet Peninsula, on the east coast of the state;

public access to reserved land on Mount Wellington, behind Hobart, was being denied; and it was feared logging would take place in a forest reserve on the Gordon River on the west coast.[11] Several influential Tasmanians were becoming concerned about the state's reserves, including Henry Dobson, a politician and founding president of the Tasmanian Tourist Association; William Crooke, a major figure in the Southern Tasmanian Railway Association and the Labor League; and Evelyn Emmett, the Director of the Tasmanian Government Tourist Bureau.[12]

The opportunity to change the reserve system came in 1914 when Tasmania's first significant Labor government was elected. (Labor had an earlier short period in government in 1909.) Later state Labor governments showed little interest in conservation, and were generally ardent backers of development of the state's natural resources, but the 1914 government was somewhat receptive to preservation ideas. The year after its election it passed the *Scenery Preservation Act*, the most progressive legislation of its type in Australia at the time.[13] The act established a Scenery Preservation Board that would recommend new reserves and manage existing ones. Although a major legislative step forward, the act fell short of comprehensive conservation legislation. There was little doubt the emphasis of both the act, and the board, was on scenery and tourism management rather than the protection of natural areas. A conspicuous hole in the *Scenery Preservation Act* was its inability to create reserves to protect animals; it would be another thirteen years before any such legislation was enacted. Summing up the priorities of the legislation and the new board, the first chairman of the Scenery Preservation Board stressed that the areas of highest reservation priority would be waterfalls, gorges, rocky outcrops, 'commanding viewpoints' and 'other places of historical or scenic interest'.[14] Comprehensive conservation did not get a look in. This was particularly the case after the board came to include representatives from the Hydro Electric Commission (HEC), the Mines Department and the Forestry Commission, who were concerned to ensure it did not reserve land their agencies had their eyes on.

Although Tasmania was the first state to enact legislation like the *Scenery Preservation Act,* it was the last state to create a national park reserve.[15] The term 'national park' came out of the United States; US artist and explorer George Catlin is credited with first using the term in the 1830s.[16] Catlin was troubled by the fact that the frontier of the United States was fast disappearing. He unsuccessfully proposed that the entire area west of the Mississippi River be preserved to save the land as well as the culture of its native inhabitants.[17] Having created the term, the US also established the world's first national park—Yellowstone, established in 1872 (although a small part of present-day Yosemite National Park was preserved eight years before as a state park, not a national park). Tellingly, the US Congress only passed the park's authorising legislation after being satisfied it did not have any other economic purpose.[18] Other North American national parks soon followed, including Banff, in Canada, created in 1885, and Yosemite which was formally reserved as a national park in 1890. As with Tasmania's first reserves, tourism was a major factor behind the creation of most of the early North American national parks. The declaration of Yellowstone National Park was partly the result of lobbying by the Northern Pacific Railroad Company, which provided transport to the park.[19] Australia was also an early starter in the establishment of national parks with its first national park, today's Royal National Park south of Sydney, proclaimed in 1879.

For some time it looked as though Mount Wellington, the large cliff-lined mountain range behind Hobart, would get the honour of becoming Tasmania's first national park as a consequence of public pressure, led by the Tasmanian Tourist Association. Trips to the mountain became increasingly popular, with the number of visitors quadrupling between 1902 and 1907.[20] Legislation was passed in 1906 to reserve part of the eastern slopes of the mountain,[21] but unhappily for those hoping to make the mountain the state's first national park there was strong opposition to the reservation of the mountain on the grounds that it might compromise the quality of the city's water supply, which was mainly drawn from its slopes.[22] Hobart Council had considerable public support for this position, sufficient to ensure the national park was not declared.

Tasmania's first national parks

None of Tasmania's reserves officially became national parks until 1946 when a number were formally renamed as national parks to increase their tourist appeal.[23] Despite this, land reserved at Mount Field was known as 'National Park' before 1946 (but was only officially renamed Mount Field National Park in 1947). Both it, and reserved land on the Freycinet Peninsula on the state's east coast, would end up becoming the first reserves that would go on to become formally recognised national parks in Tasmania.

The idea of creating a major reserve at Mount Field is attributed to Leonard Rodway, a leading botanist and major figure in the Tasmanian Field Naturalists' Club, as well as to Herbert Nicholls, a prominent Tasmanian politician and member of the Tasmanian Tourism Association.[24] Nicholls and Rodway's idea was taken up by William Crooke, who in 1911 organised a visit to the region by parliamentarians, including premier Elliot Lewis. Two years later Lewis, who was by then no longer premier, led a delegation to the lands minister to promote the idea of a Mount Field reserve.[25] He met with some success, but the government of the day generally agreed only to create a small reserve of about 2200 hectares. Following the election of the new Labor government the following year, however, the reserve was expanded to nearly 11 000 hectares, exceeding even the expectations of the park's proponents.[26]

The opening of the new Mount Field reserve left no doubt that the Tasmanian government mainly viewed the reserve as an economic development opportunity. It also left no doubt that there was a fierce tension between development and conservation, even in those days. Woodchopping competitions were held as part of the opening ceremonies. As part of the opening speeches, Henry Dobson emphasised the potential of the new park to boost tourism in the area. William Crooke hit back at Dobson, saying: 'the idea of the Park was not originally conceived simply for tourists'.[27]

Unbeknownst to Dobson and Crooke, their differing philosophies would be a recurring theme of Tasmanian reserve management well into the future. The clash between tourism and nature conservation is *the* tension that pervades the management of Tasmania's national parks. It is no closer to being resolved today than it was in Crooke and Dobson's time. In fact ever-increasing tourist numbers have made it an even greater issue. There does not have to be such a clash, provided prudent and sensitive management practices are employed. All too often, however, both the creation and the management of Tasmania's reserves have been victim to indiscriminate and unthinking political decisions that have paid scant regard to environmental sustainability. National parks have become defining symbols of Tasmania and in many ways embody what is best about the state. But like the state's rivers and trees, they also embody what is worst about it. All too often Tasmania's national parks have been caught between development and preservation pressures and have generally lost out.

Freycinet Peninsula was proclaimed a reserve at the same time as Mount Field. The area had originally been recommended as a reserve in 1894 by the Australian Association for the Advancement of Science, though the government of the day rejected the proposal in order to protect local farming and mining interests. A bit over a decade later, however, part of the peninsula was reserved under the *Game Protection Act,* mainly for fauna conservation reasons, although the lack of any ranger in the area meant significant hunting in the reserve continued.[28] After the initial reservation, supporters of greater protection of the peninsula continued to stress its recreational and tourism potential, along with its worth as an area of fauna protection. In 1916 they succeeded in having it proclaimed a 10 000 hectare reserve, roughly the same size as Mount Field.

Creation of the Cradle Mountain–Lake St Clair reserve

After Mount Field and Freycinet, the next reserve created in Tasmania that would eventually become a national park was the Cradle Mountain–Lake St Clair reserve. Both Cradle Mountain and Lake St Clair are icons of Tasmania's wilderness and the area's addition to the state's reserve system was a major step forward for wilderness protection. Its significance was reinforced by the fact that the original area reserved was much larger than either the Mount Field or Freycinet reserves, about six times their size. The main campaigner for the northern part of the reserve was Gustav Weindorfer, who became something of a local folk hero. In 1921 he travelled to Launceston and Hobart to lobby politicians, newspaper proprietors and the Scenery Preservation Board about creating a reserve at Cradle Mountain.[29] The following year his dream came true.

Weindorfer's empathy for Cradle Mountain was a product of his childhood. Born in the Austrian alps in 1874, he emigrated to Australia in 1900. In 1906 he married Kate Cowle, who shared his interest in the environment. They soon moved from Victoria to Kindred, near Cradle Mountain.[30] In 1909 the Weindorfers made their first visit to Cradle Mountain, climbing it in January 1910. They were spellbound by the summit view. According to a fellow climber, upon reaching the top Weindorfer said: 'This must be a national park for the people for all time. It is magnificent and people must know about it and enjoy it.'[31]

The following year the Weindorfers bought land at Cradle Mountain and in 1912 began building a chalet called *Waldheim* ('forest home'), a replica of which still stands today. After Kate's death in 1916, Weindorfer made Cradle Mountain his permanent home. Following the upgrading of the access road to the region after the First World War, he energetically promoted Cradle Mountain through lectures augmented by lantern-lit pictures. His enthusiastic advocacy made him one of the state's first conservationists.

Soon after Weindorfer's 1921 lobbying, a parliamentary delegation visited the area. Although interested in the reserve's creation, the board was concerned about the potential for such a large reserve to lock up the timber, grazing and mineral resources of the area.

Tension about the withdrawal of access to the resources of the Cradle Mountain–Lake St Clair reserve was the first occasion when the Scenery Preservation Board was caught between the competing pressures of conservation and development. In later decades the board's response to resource pressures defined its character and ultimately led to its demise. The future of the resources of the Cradle Mountain–Lake St Clair area was a crucial early test of its conservation resolve but its response was wanting.

The Scenery Preservation Board agreed to create a Cradle Mountain–Lake St Clair reserve but said its creation should not hinder exploitation of the area's resources. The chairman of the Board, E.A. Counsel, said the reserve could not be proclaimed under the provisions of the *Scenery Preservation Act* (which precluded use of natural resources once a reserve was established). He argued that prohibiting the exploitation of the area's resources would generate a lot of opposition to the reserve.[32] Counsel and the board therefore recommended a major watering down of the *Scenery Preservation Act* that would allow resource exploitation even after an area was reserved. An amendment to the act was passed in 1921 before the Cradle Mountain–Lake St Clair reserve was proclaimed in 1922. This was the first big cave-in by Tasmania's reserve managers to development interests. There were many more to follow but it was an unfortunate early indicator that development would generally triumph over reservation. In most peoples' eyes, development equalled money: reserved land did not; it was that simple. Time and time again, the state's reserve management authority would be pressured to put development first.

Further development pressure on the Scenery Preservation Board

Given the precedent set by the 1921 amendment to the *Scenery Preservation Act,* when minerals were found in the Cradle Mountain–Lake St Clair reserve in the late 1930s, it would have been reasonably expected that the board would mount little resistance to their exploitation, particularly since the state government was keen to see the mining go ahead. But, surprisingly, the board found its spine and put up a stiff opposition. It managed to outmanoeuvre several attempts to allow the mine to proceed. Despite increasing government pressure, the board stood its ground.

In 1939, the board finally lost its battle but only after the government further watered down the *Scenery Preservation Act,* making it quite clear that parts of the act could be suspended to allow mining to take place in a reserve.[33] In the end, the act was suspended over 1200 hectares of the reserve around the mine. The mining activity only lasted for five years, however, and the area (at Pelion in the middle of the reserve) was eventually returned to the national park. Although it seems bizarre today that a mine could be allowed in the middle of a national park, the idea was not so foreign at the time. The Canadian *National Parks Act,* for instance, allowed both logging and mining within national parks until 1930.

At much the same time, the Scenery Preservation Board came under pressure to allow another type of development at the southern end of the Cradle Mount–Lake St Clair reserve, at Lake St Clair. During the 1930s the HEC was expanding its operations in central Tasmania, where Lake St Clair is situated. Part of the

HEC's expansion involved developing the lake through the construction of a small weir at its outlet. The weir raised the lake's level by 3 metres, sufficient to flood a picturesque beach, known as the Frankland beach, on its southern shore as well as some equally appealing small islands. When news of the inundation became public, it sparked one of the first instances of public indignation about development in Tasmania's reserves. Not wanting to repeat the experience with the proposed mine in the Cradle Mountain–Lake St Clair reserve nor to bow to public pressure, in 1940 the HEC simply presented the lake's flooding as a fait accompli to the board, which never formally gave its endorsement.

The Cradle Mountain–Lake St Clair reserve threw up yet another contentious development proposal in the late 1930s. Between 1934 and 1939 the Labor state government had spent a lot of money upgrading the road into Cradle Mountain. It had also developed a number of other major tourist roads that were part of a broader triumvirate of Great Depression development initiatives in tourism, hydro development and forestry. In 1938 the Cradle Mountain Reserve Board endorsed the idea of extending the road through the middle of the Cradle Mountain–Lake St Clair reserve to Lake St Clair. The war intervened, and when the idea was revived in 1944 there was more opposition than support, even within the Cradle Mountain Reserve Board, so the road did not go ahead.[34]

The Cradle Mountain–Lake St Clair reserve mine proposal and the flooding of Lake St Clair were curtain-raisers for the Scenery Preservation Board's biggest tussle over a reserve development proposal—its clash with Australian Newsprint Mills (ANM) over the company's proposal to excise part of the Mount Field National Park. That tussle permanently scarred the Scenery Preservation Board and led to the creation of Tasmania's first conservation organisation, the Tasmanian Flora and Fauna Conservation Committee. The board never fought a development proposal for a reserve again.

ANM first approached the Scenery Preservation Board about taking some of the prime forests of Mount Field National Park for its papermaking operations in 1946. The company was interested in the tall eucalypts that ran along the park's western boundary and formed part of the Florentine Valley, which once hosted Tasmania's most extensive forest of tall eucalypts. ANM, and the supportive Tasmanian government, variously argued that it had always been intended that the forests be included in the company's resource allocation and that the company faced eventual closure without the extra resource (a time-honoured argument of many Tasmanian development interests). ANM also claimed there was doubt about the true boundary of the national park and that the forests it wanted were not accessible to the public so should not have been in the national park in the first place. The company ended up making three separate approaches to the Scenery Preservation Board, suggesting various schemes through which it would exchange the prized Florentine Valley forests for other, lesser, ANM-allocated forests that bordered another part of the national park. The Scenery Preservation Board resisted each approach. The board argued the forests ANM wanted formed 'a most pleasing foreground for the lovely panorama' that could be had from the Mount Field reserve. It also said it was important to preserve one of the last 'virgin' stands of the mighty *Eucalyptus regnans* that still existed in the state.[35]

The board was united in its opposition. Even HEC head Alan Knight, who was a member of the board, argued that 'those areas are put in our trust and we would be going beyond our proper function in handing the area over'.[36] But the board was fighting an impossible battle. In 1947 Premier Robert Cosgrove threw his weight behind ANM's claim and after the board resisted the third approach from the company, the government in 1950 passed the *National Park and Florentine Valley Act*. The act revoked 1523 hectares of national park forest, substituting it for 1552 hectares of forest along its southern boundary.

Although ultimately vanquished, the Scenery Preservation Board showed a resolve that is rare amongst Tasmanian government agencies. It displayed enormous courage but never recovered from the fight and was acquiescent on later reserve development issues. Although not well remembered today, the conflict was the first major fight over a national park revocation in Australia. One of the saddest legacies of that fight is that precious few of Tasmania's original tall trees, particularly its *Eucalyptus regnans*, remain. Of about 100 000 hectares of the species that existed when Europeans first arrived on the island, only about 13 000 hectares still stand today.[37]

The 1930s to the 1960s

Between the 1930s and the 1960s the Scenery Preservation Board made four more reserves outside south-west Tasmania that later became national parks. The first was in the Hartz Mountains area, south of Hobart. The original impetus to create the park was a state government proposal to build a road to the area. In 1938 the Public Works Department proposed that the area be made a reserve, partly to justify the spending of the then extraordinary amount of £30 000 on the road's construction.[38] The idea of the reserve had popular support although there was no question in most peoples' minds that it should primarily be created for tourism development and not conservation. The Hobart *Mercury* commented: 'Hartz makes a particular appeal to tourists . . . at present, however, it is accessible only by packhorse or by foot. It is an area calling loudly for development.'[39] Creation of the reserve was uncontroversial. It was proclaimed in 1939, although parliament ended up blocking the building of the road on cost grounds and it was not finally built until several decades later.

The next major reserve that would end up becoming a national park was at Frenchmans Cap, on the northern edge of the state's south-west wilderness. Frenchmans Cap, a distinctive peak with a sheer 400 metre quartzite rock face that falls from its summit to the valley below, is an icon of western Tasmania. The view of 'the Cap' became well known after the first road link between eastern and western Tasmania was completed in the early 1930s, and the area was particularly well known amongst bushwalkers. In 1940 a walker, Ray Livingstone, wrote to the Scenery Preservation Board proposing that it reserve the area around the mountain. As if to confirm its focus on scenery rather than conservation, the board at first thought the region was too remote to be made a reserve. But it persuaded the government to build a walking track to the mountain then consented to the new reserve, which was proclaimed in 1941.[40]

Outdoor enthusiasts of a different kind were behind the proclamation of a national park in the alpine area of Ben Lomond in the state's north-east. After the First World War, skiing became popular in Tasmania, leading to the establishment of the Northern Tasmanian Alpine Club in 1929. The club investigated several areas that might be suitable for ski development throughout the north of the state before settling on the Ben Lomond area, where it developed access routes and modest accommodation throughout the 1930s.[41] After the Second World War, the Scenery Preservation Board became interested in reserving the Ben Lomond plateau but the club resisted the idea, thinking it would lose control of its ski grounds. It lobbied for the reservation plan to be abandoned but the board was unsympathetic and proclaimed the reservation in 1947.[42] Mount Barrow, near Ben Lomond, was also declared a 459-hectare national park but eventually this status was revoked because it was deemed too small.[43] The clash between the Northern Tasmanian Alpine Club and the Scenery Preservation Board was the first instance where people who saw themselves as traditional users tried to thwart the creation of a national park in Tasmania. Along with developers, those who claimed to be traditional users, such as graziers and horse riders, frequently became objectors to the creation of national parks.

The Scenery Preservation Board also identified an area at Rocky Cape on the north-west coast as a national park. A small reserve of 81 hectares had been proclaimed in the area in 1912, mainly to preserve the unique stands of *Banksia serrata* that grow there. After the reservation, the Burnie Field Naturalists' Club kept an active interest in the area and agitated for a larger reserve.[44] In 1965 a proposal was put to the board to proclaim a reserve of about 3000 hectares but it was equivocal and deferred a decision before finally proclaiming a reserve of about 1600 hectares in 1967.[45]

Yet another national park created by the board was in the south-west corner of Flinders Island, off the north-east corner of the Tasmanian mainland. The park mainly covered the Peaks of Flinders range that rises to nearly 800 metres. Proclaimed in 1967, it was given the name of Strzelecki National Park in 1972 in honour of Count Paul Strzelecki, a Polish scientist and explorer who had climbed many of the island's peaks in 1842.[46]

Of the five national park reserves created by the Scenery Preservation Board outside the south-west of the state between the 1930s and 1960s, only Rocky Cape and Strzelecki could be said to be largely created for conservation reasons. As its name made all too clear, the Scenery Preservation Board was created to preserve scenery and it rarely wavered from making that its priority. Although the first half of the twentieth century saw significant growth in visitation to natural areas, this did not necessarily translate into a popular conservation ethos. The clash over ANM's bid to excise part of the Mount Field National Park saw the first flickerings of such an ethos. But it was not until the 1960s, and the fight to save Lake Pedder, that conservation became a big issue in Tasmania. And it was not until that time that national parks came to be created mainly for conservation reasons.

The Lake Pedder and Southwest national parks

Although controversial, the skirmishes experienced by the Scenery Preservation Board over the excision of parts of Mount Field and Cradle Mountain–Lake St Clair national parks were low key compared to the trauma it would experience over the creation, and destruction, of national parks in south-west Tasmania.

The wilderness of the south-west is magnificent. It is a vast rugged area punctuated by sharp, remote ranges and peaks that pierce the sky. It contains rock from almost every geological era, and a wealth of plant species and vegetation types, ranging from forest through to buttongrass moorlands and alpine meadows. It is recognised as an International Centre for Plant Diversity by the World Conservation Union. Its fauna is characterised by an unusually high proportion of endemic species, some with links to ancient predecessors.[47] South-west Tasmania is one of the few wilderness areas in Australia where it is still possible to stand on a peak and see no signs of human activity. Indeed, it is possible to walk in some parts of the south-west and see nobody for days. Until well after the Second World War the area was known as 'the empty quarter' and only first became significantly explored by bushwalkers in the 1920s. The highest peak in that part of the state, Mount Anne, was first climbed by non-Aboriginal people in 1929. South-west Tasmania is one of the great wilderness areas of the world and it was for good reason that it became the battleground for some of Tasmania's most dramatic wilderness clashes.

The debate about the reservation of the south-west began quite innocently. Although some isolated parts of the area had early reservation (such as the Gordon River), the first proposal for any major type of reservation came, remarkably, from the Country Women's Association (CWA). In 1947, partly motivated by concern about hunting dogs that had been abandoned in the south-west, the organisation proposed to the Scenery Preservation Board that part of the area be reserved. The board was unimpressed, however, and rejected the CWA's proposal.[48] In the same year, bushwalking parties began to organise food drops into the area from light planes, which also began using the beach of Lake Pedder as an airstrip. These developments increased the accessibility of the south-west and the interest of bushwalkers.[49] Back then the nearest road to the region finished at Maydena, a town some 40 kilometres away from Lake Pedder, so the enhanced access made a considerable difference.

But bushwalkers were not the only new visitors to the area. At the same time the HEC began placing flow meters on the upper reaches of the Gordon River, which flowed near Lake Pedder. This raised the suspicions of the walkers, and prompted the Hobart Walking Club to propose to the Scenery Preservation Board that Lake Pedder be made a national park. Upon receiving the proposal, the board viewed slides of the area before referring the proposal to the HEC, the Forestry Commission and the Department of Mines.[50] Remarkably, the HEC raised no objection to the national park. It even noted that it had no development planned for the area for at least twenty years (although it did ensure that the valley of the lower Serpentine River, on which the dam that would eventually flood Lake Pedder was situated, was excluded from the park).[51] The proposal seemed uncontroversial and in March 1955 the board declared a 25 899 hectare national park with slightly different boundaries to that proposed by the Hobart Walking Club. The Lake Pedder National Park remained the only national park in the south-west for the next thirteen years.

The creation of the Lake Pedder National Park coincided with a major change in the leadership of the Scenery Preservation Board. From 1938 to 1953 the board had been chaired by the energetic Surveyor-General of Tasmania, Colin Pitt. He led the fights against mining and hydro electric development in Cradle Mountain–Lake St Clair National Park as well as against ANM's forestry excision from Mount Field National Park. Pitt was always keen to see new reserves created throughout the state and provided valuable leadership to the board. Following Pitt's death a new Surveyor-General,

Frank Miles, took over but he was much less enthusiastic about his position on the board.[52] Miles's elevation saw the board lose its assertiveness and it never again found the fighting spirit it had under Pitt.

Buoyed by its success with the Lake Pedder National Park, the Hobart Walking Club went back to the board in 1957 with a proposal that the stunning Arthur Range, near Lake Pedder, also be reserved. But the board was not enthusiastic to extend the reservation of the south-west and the proposal was knocked back.[53]

Despite the board's reluctance, more proposals for broader reservation of the south-west kept coming. A proposal for an enhanced south-west national park was referred by the board to the Mines Department, Forestry Commission and HEC, all of whom opposed the idea. The official line from the state government was that the concept had merit but was premature.[54]

In 1958 the HEC installed another flow recorder on the Gordon River, and in 1962 built a jeep track into the Serpentine River tributary of the Gordon before installing a flow meter on that river too. Despite this, the HEC kept to the line used when the Lake Pedder National Park was created, that no imminent development was planned. The HEC's head, Alan Knight, even said: 'the possibility of power development in this area in the foreseeable future is remote'.[55]

The veil on the HEC's real intent was finally lifted on 10 December 1962 when pro hydro Premier Eric Reece announced that the Gordon River would be investigated for its hydro electric potential. It would be another four and a half years, however, before the HEC's plans for Lake Pedder, and for the south-west in general, became clear. But it was obvious the area was under threat in a way it had never been before. It was also obvious that neither the HEC, nor the state government, were sympathetic to any meaningful reservation of the south-west. Reece said of the land to be flooded by a dam on the Gordon River that it '. . . will be mostly buttongrass plains which are now waste areas'. Knight claimed 'similar hydro developments in the Central Plateau areas had not affected the scenic beauty of the area, rather they had opened them up with first class access roads'.[56]

In the same year that Reece announced the Gordon River investigations, Ron Brown, the Deputy President of the Tasmanian upper house, put up his own south-west national park proposal. The Reece government dismissed his proposal so in November 1962 Brown took charge of the first of five major conservation organisations that would be formed to protect the south-west—the South West Committee.[57] The committee lodged a formal proposal for a south-west national park with the Scenery Preservation Board in 1964. The board was reluctant to be the only government agency to recommend such a new national park, so it suggested that an interdepartmental committee be formed to evaluate the proposal.[58] This course suited Reece, as it would divert attention from the HEC's plans for Lake Pedder. In 1965 he announced the setting up of the interdepartmental committee but made it clear the South West Committee was not to be amongst its members.

The interdepartmental committee gave its findings to the Minister for Lands in April 1967, recommending the creation of a new national park that would extend from the flooded Lake Pedder through to the south coast of the state. By implication, it backed the HEC's plans and carefully avoided recommending reservation of any areas that might have significant mineral or forestry resources. A representative from the Scenery Preservation Board was on the committee and was a party to its compliant decision. The board's earlier defiance of mining, forestry and hydro development in the 1930s and 1940s was by then a distant memory; it was no longer prepared to rock the boat. There was no longer a voice within the government bureaucracy that was prepared to stand up for conservation. In 1966 the board oversaw the establishment of a 661 980-hectare fauna district across the south-west but this gave the area no meaningful protection. Then in October 1968 it proclaimed a new, compromised 195 867-hectare Southwest National Park.

Although the national park represented the first major reservation of the south-west, it was a disappointment to the South West Committee and all who had campaigned for meaningful protection of the area. The national park was only a third of the area proposed by the committee and was basically a token corridor of protected land. It had a convoluted boundary that carefully avoided any resources of potential mining, forestry or hydro importance. While it included some spectacular peaks and many buttongrass plains, its boundary excluded the extensive forests that ran along the eastern boundary of the south-west, the tin-bearing area around Melaleuca to its west, the limestone-bearing Precipitous Bluff region to its south-east and any significant reservation of the rivers that flowed to its west. The interdepartmental committee deliberately cut large sections off the original national park proposal to avoid controversy. Although the new park was about eight times the size of the Lake Pedder National Park, it left as many reservation gaps as it filled. It was a very small step forward for the south-west.

The end of the Scenery Preservation Board

After the HEC's report on its Lake Pedder scheme was tabled in May 1967, the Tasmanian upper house began an inquiry into the development. Amongst other things the inquiry looked at the scheme's approval process, including the revocation of Lake Pedder National Park. The parliamentary committee's report was critical of the role played by the Scenery Preservation Board and recommended that the board be scrapped. The conservation movement was also critical of the board's meek acceptance of the Lake Pedder National Park revocation and echoed the recommendation, arguing that the board had insufficient staff, insufficient money and insufficient muscle.

The government's initial reaction was to contemplate changes to the *Scenery Preservation Act* that would give the board greater powers but the conservation movement argued the board was unreformable. Initially, the board was consulted about possible changes but the widespread contempt it was held in after the Lake Pedder controversy ensured it was eventually ignored.[59]

Momentum for a major change of Tasmania's reserve management system increased and in 1968 a new *National Parks and Conservation Act* was put before parliament but failed to pass before state elections were called the following year.[60] The Scenery Preservation Board was resentful about its marginalisation but was unable to stop the tide that was rising against it. When created in 1916 the board, and the act it implemented, were ahead of their time but by the late 1960s both were dated and wanting.

By the time of the 1969 election all sides of politics accepted the need for major reform of the state's reserve system, although there was disagreement about what form it should take. It had also been 34 years since the Liberal Party had been in government in Tasmania and it was hungry for power. In his election policy speech, Liberal leader Angus Bethune responded to the growing public demand for better reserve and environmental management by promising that if elected, his government would introduce air pollution legislation and establish a properly resourced National Parks and Wildlife Service. The Labor Party under Eric Reece was more reticent. Reece promised he would create a new tourism ministry that would incorporate the Scenery Preservation Board. The board would still manage the state's national parks but it would be subject to the ultimate power of the Director-General of Tourism.[61] Like so many of his proposals, Reece could only see reserve management in resource terms. It had to be part of an industry and it had to make money to be worthwhile.

In the end the Labor and Liberal parties each won seventeen seats in the 35-member lower house, with the balance of power going to a Centre party member, Kevin Lyons, who sided with the

Liberals. Like Bethune, Lyons was committed to reserve management reform so in 1970 a new *National Parks and Wildlife Act* was passed and in 1971 a professional National Parks and Wildlife Service commenced operation in Tasmania.

The Scenery Preservation Board met for the last time in October 1971 to close an eventful chapter in the history of Tasmania's reserves. The positive legacy of its tenure was a reserve system that, although far from comprehensive, nonetheless covered some significant parts of the state. By 1971 the state had national parks in the south-west, on the Freycinet Peninsula, at Mount Field, at Frenchmans Cap, at Hartz Mountains, at Rocky Cape, at Ben Lomond, on Flinders Island and between Cradle Mountain and Lake St Clair. It also had major reserves along the Pieman and Gordon rivers on the west coast, along the Lyell Highway and at Mount Barrow.[62] The board's negative legacy was one of trying to manage the state's reserves on the cheap and of a preparedness to support the government's attempts to revoke parts of reserves if they conflicted with development. By 1971 the board had been party to two major excisions from Cradle Mountain–Lake St Clair National Park for mining and hydro development; two excisions from Freycinet National Park in 1941 and 1942 for mining; the ANM excision of 1489 hectares from Mount Field National Park for forestry development; the revocation of the Lake Pedder National Park; and 1902 hectares of excisions from Hartz Mountains National Park in 1943, 1952 and 1958 for forestry.[63] Only two of these revocations had been resisted by the board.

By the early 1970s Tasmania's national parks had had a patchy history. They were on the map but all too often had been compromised by poor management and a preparedness to sacrifice them to development interests. The state lacked a representative reserve system but it had a new, professional organisation to manage its reserves. Expectations among conservationists were high that the National Parks and Wildlife Service would deliver a brighter future.

IO

The National Parks and Wildlife Service

The establishment of the National Parks and Wildlife Service (NPWS) was an initiative full of promise. Finally, Tasmania had a well-resourced and independent conservation and land management agency that would provide a much-needed environmental voice within the state government bureaucracy. Once the *National Parks and Wildlife Act* was passed in 1970, moves were quickly afoot to get the NPWS up and running. In April 1971 the first director of the Service, Peter Murrell, was appointed with an initial staff of 59. Murrell had been a senior manager of the New South Wales Parks and Wildlife Service and had also been Assistant Conservator of Forests in Kenya. He was keen to reinject a fighting spirit into the management of the state's reserve system. Murrell did not seem to want to shy away from a fight. He even publicly declared: 'I think it is inevitable there will be conflicts between our authority and others such as the Hydro Electric Commission'.

New national parks outside the south-west

Before the creation of the NPWS, in addition to the Scenery Preservation Board, there had been a separate Animal and Birds Protection Board which became known as the 'Fauna Board'. As its name suggested, the Fauna Board was dedicated to the protection of the state's wildlife and tended to be more proactive than the Scenery Preservation Board. In the late 1950s the Fauna Board shared a concern with the nascent conservation movement that the state's reserves were not comprehensive enough to adequately protect wildlife habitats. A review of the reserves in the late 1950s concluded there were too many small wildlife sanctuaries throughout the state and that larger reserves were needed for adequate fauna protection.[1] In response, the Fauna Board created a new category of reserve—the fauna reserve—and began searching the state's east coast for a large area of land that could become a dedicated wildlife sanctuary. The idea of creating such a sanctuary was partly inspired by their apparent success in Africa.[2] The National Parks and Wildlife Service has always administered a range of reserves that had less protection than national parks. Today these include nature reserves, state reserves, historic sites, game reserves and regional reserves.

The Fauna Board identified Maria Island, situated 5 kilometres off the Tasmanian mainland north-east of Hobart, as a suitable site. From 1964 the state government began buying farmland on the island as it came up for sale and in 1967 the Fauna Board started stationing rangers on the island. Between 1969 and 1971 the board imported threatened animals to the island, in a regular Noah's Ark style operation. These included: Forester kangaroos, Bennett's wallabies, Flinders Island wombats, brush and ring-tailed possums, potoroos, bettongs, echidnas, marsupial mice, bandicoots, and pademelons.[3] When the island was made a fauna reserve in 1965 the Federation of Field Naturalists of Tasmania had suggested it be made a national park but the Scenery Preservation Board rejected the idea. The new NPWS, however, took a different view and in 1972 managed to get the island proclaimed a national park.

A similar concern about preserving wildlife was the major force behind the creation of another national park the year after Maria Island: the Mount William National Park, on the north-east coast. Not long after the Fauna Board settled on Maria Island, the Tasmanian Farmers and Graziers' Association approached the state government about the extent of land clearing in the north-east. It was particularly concerned about the threat the clearing posed to the state's population of Forester kangaroos. The Forester kangaroo is Tasmania's largest marsupial and the only kangaroo species found on the island. By the 1960s it was experiencing significant decline in both its numbers and range.[4] Its range was being reduced by land clearance, an issue that was heightened by the ownership and clearance of over 100 000 hectares of land in the north-east by the British Tobacco Company for real estate development. The Fauna Board took up the issue and proposed a new fauna reserve for the region that would include some British Tobacco land.

In 1970, shortly before the board's functions were taken over by the NPWS, it resubmitted plans to the government for the creation of a large reserve. Some of the reserve land would need to be compulsorily acquired from British Tobacco. The state's Surveyor-General supported the board's plans and the NPWS eagerly continued its vision. At the same time, questions were being asked about how British Tobacco had acquired the land. In 1971 a court hearing was held into allegations that in order to get the land the company had fraudulently conspired with the Reece Labor government, which had lost office in 1969. Although the hearing was dismissed because of insufficient evidence, the ill feeling over the issue probably influenced the new Liberal government in supporting the NPWS's proposal to resume some of the company's land. Finally, in October 1973, the new 8640-hectare Mount William National Park was proclaimed.

Mount William National Park was the first national park in Tasmania whose creation caused a clash between pre-existing owners and the NPWS. Similar tensions were experienced when the NPWS moved to create a national park in the Asbestos Range area on the mid north coast. Along with the area around Rocky Cape National Park, Asbestos Range included some of the last undeveloped coastline along the north coast. As early as 1938 the local Beaconsfield Council had suggested creating a national park in the region.[5] The Mines Department opposed the idea and it was forgotten about until the early 1970s when interest began to be shown in developing land around the Asbestos Range. In 1971 the Tasmanian Town and Country Planning Commissioner courageously refused permission, on several occasions, to a Sydney developer who wanted to subdivide a farm that was coming up for sale in the area. The NPWS outflanked the developer by buying the land using funds provided by the Whitlam government through its National Estate grants program.

Unfortunately the NPWS lacked the funds to buy land that could form the eastern part of the national park. This land was bought up by another developer and clearing began. The NPWS then took the bold step, again, of convincing its minister to compulsorily acquire part of the developer's land. This created a backlash from local residents and the Beaconsfield Council, both of whom feared the NPWS would stop further development and halt some of the traditional uses of the area, such as horse riding. The government backed the NPWS with the result that in 1976 the Asbestos Range National Park was proclaimed.[6] In 2000 the park was renamed Narawntapu National Park to avoid confusion with the abestosis disease and to acknowledge its Aboriginal heritage.

Conflict of a different kind surrounded the creation of the Walls of Jerusalem National Park, in the state's Central Plateau region, three years later. The Central Plateau, particularly its western reaches, is a stunningly beautiful elevated area covering about 5000 square kilometres that is home to extensive alpine environments. It features meadows of native grass and stands of alpine eucalypt and pines, punctuated by rocky outcrops. It also contains thousands of lakes: more than 4000 large and small lakes are dotted throughout the region.[7] It is the only expression of this type of country in Australia. As early as 1946 an outdoor enthusiast, Jack Thwaites (who was a member of the Scenery Preservation Board), proposed to the board that the Walls of Jerusalem be made a national park. The board rejected the idea, thinking the area was too isolated,[8] just as it had initially reacted to the Frenchmans Cap National Park proposal. But in 1962 it reconsidered the idea after another proposal was submitted by the Launceston Walking Club. This time, the board referred the proposal to the Forestry Commission and the Hydro Electric Commission (much as it had done with the first proposal for a south-west national park). The Hydro Electric Commission replied that it planned to reserve all of the Central Plateau above 914 metres (3000 feet) for water catchment purposes. The Scenery Preservation Board thought this was an adequate alternative way of protecting the area and again rejected the national park idea.[9]

At the time of the second rejection, however, concern was mounting about the impact that sheep grazing and the spread of rabbits and hares was having on the plateau. Sheep had long been raised in the region and by the 1950s as many as 160 000 were grazing on its delicate alpine environment between December and May each year. Timber harvesting in some of the lower reaches of the plateau, three hydro power stations on its rivers and extensive populations of trout throughout its waterways placed additional pressures on the plateau's ecosystem. In response to mounting concern, in 1976 the Lands Department produced a draft management plan that mainly focused on the western and northern parts of the plateau. The final management plan, published in 1981, was little different to the draft and resulted in the creation of a Protected Area which in fact gave the area little meaningful protection. The pressure for proper protection did not relent, however, and in 1981 a 11 510-hectare Walls of Jerusalem National Park was finally proclaimed.[10]

Enlargement of the Southwest National Park

The original Southwest National Park proclaimed in 1968 was widely viewed by the conservation movement as severely compromised. From 1973 onwards alternative reservation plans were released by the South West Committee, the Australian Conservation

Foundation, the Tasmanian Conservation Trust, and the South West Action Committee, as well as various outdoor clubs. The rush of grander reservation plans was stimulated by the anticipated release by the NPWS of its new draft south-west management plan.[11] Although the NPWS's draft plan was leaked, its official release was delayed until 1975 to allow for resolution of a court case the Tasmanian Conservation Trust was bringing against a Melbourne company, Mineral Holdings (see Chapter 8). The company wanted to mine limestone at Precipitous Bluff, then situated outside the eastern boundary of the Southwest National Park.

The draft plan was the subject of many battles between the NPWS and the Mines Department, the Forestry Commission and the Hydro Electric Commission. Although the plan recommended some additions to the park (including Precipitous Bluff, the South Coast Range and areas north of the Port Davey–Bathurst Harbour area) it still fell well short of the conservation movement's expectations. The Australian Conservation Foundation was particularly scathing, declaring that 'the contents of the document reveal complete subservience by its author(s) to the interests of commercial exploitation.'[12]

The NPWS's plan was no more acceptable to the state government than it was to the conservation movement, so the government appointed a three-person advisory committee, headed by University of Tasmania Chancellor Sir George Cartland, to re-examine the issue. The Cartland committee sought public submissions and released a preliminary report in mid 1976. The composition of the committee seemed biased in favour of development interests so the South West Action Committee boycotted its deliberations.[13]

Like the NPWS's draft plan, the committee's preliminary report was a disappointment, again failing to advocate adequate reservation of Tasmania's greatest wilderness area. Although the committee noted that 'the area is of world heritage status and is a unique natural asset', it nonetheless concluded that 'it would be impracticable and unwise to constitute the whole of south-west

Tasmania as a national park'.[14] The Cartland Committee endorsed the controversial idea of swapping part of Hartz Mountains National Park for the forests that would be reserved in the expanded Southwest National Park around Precipitous Bluff. It also largely agreed with the compromised new national park boundaries advocated in the NPWS's draft plan. The government therefore added 206 813 hectares to the Southwest National Park in November 1976, doubling its area to 399 440 hectares.

The Cartland Committee's deliberations coincided with the election of the new Liberal–National Country Party federal government headed by Malcolm Fraser. During his election campaign Fraser promised to establish a world-class national park in south-west Tasmania and to fund a survey of the region's resources. The results of the survey were incorporated into the committee's final report, released in late 1978, which did not recommend any changes to the national park boundaries.[15]

The NPWS made it clear that, like the conservation movement, it considered the new boundaries of the national park inadequate. In 1976 the Australian Conservation Foundation presciently declared that the south-west should be listed as a World Heritage Area (the Australian government having signed the World Heritage Convention two years before).[16]

By the time of the publication of the Cartland Committee's final report it had been thirteen years since Eric Reece had announced the establishment of an interdepartmental committee to examine the reservation of south-west Tasmania, but the issue was no more resolved than it had been in the mid 1960s. The reservation of the south-west was *the* national park issue in Tasmania from 1965 through to 1990 and remains a potent issue today. The 1976 expansion of the Southwest National Park secured some areas in the park but left too many conspicuous gaps.

A bizarre incident took place outside the south-east boundary of the Southwest National Park in 1972. That year the army proposed the establishment of an artillery firing range at Cockle

Creek, in an area later incorporated into the park. The site included the start of a major track into the park, the South Coast Track. The artillery range would have resulted in unexploded shells, accidents and significant environmental damage if allowed to go ahead. The South West Committee took up the issue with gusto, quickly turning it into the 'Battle for Cockle Creek'. The idea was quietly dropped as a result of the uproar from the media and users of the area including beekeepers, bushwalkers, yachting enthusiasts and fishermen.[17]

The Wild Rivers National Park

When the Hydro Electric Commission (HEC) unveiled its plans for the scheme that would flood Lake Pedder in 1967, it was clear that an integrated hydro development was planned for the Gordon and that, eventually, the lower reaches of the Franklin River (which flows into the Gordon) would be threatened. When the proposal for Lake Pedder was announced the HEC downplayed its interest in the Franklin. It kept denying having designs on the river even when asked about test drilling it conducted in the vicinity in 1964.[18] Like the HEC's plans for Lake Pedder, there were too many coincidences for the activity around the Franklin to be without purpose. Four years before the test drilling started, part of a reserve that had been created on the lower reaches of the Gordon River was 'accidentally' revoked by the Scenery Preservation Board at a time when the head of the HEC was an influential member of the board. In the 1970s the HEC built an access road to a possible dam site on the upper reaches of the river and by 1978 had spent $6.5 million investigating dam sites on the Franklin, Gordon and nearby King River. Finally, in October 1979, the HEC came clean. It said it wanted to build a massive $366 million integrated scheme on the Franklin that would generate 296 megawatts of electricity with its first stage finished by 1990 (see Chapter 3).

Once the HEC's plans were revealed, the NPWS knew it faced its greatest political challenge yet but took an independent approach by undertaking its own evaluation of the hydro plans. Its review was critical, arguing that the HEC failed to provide an adequate evaluation of the environmental impact of the scheme. It also pointed out that the HEC had given little consideration to other electricity generation options and had failed to consider other land uses for the area, including the possibility that a national park might be created in the region. The HEC was affronted by the NPWS's evaluation; it was not used to other government agencies standing up to it. It claimed the NPWS's evaluation constituted 'a substantial attack on the integrity of the Commission and some of its individual officers', and even said it would seek legal advice about whether the NPWS's report defamed some of its staff.[19] Fortunately for the NPWS, its stance was supported by its minister, Andrew Lohrey.

In late 1977 Lohrey had been made minister for Resources and Energy—a vast portfolio that embraced forestry, the HEC and the environment—in the new Labor government of Premier Doug Lowe. Lohrey brought a discerning attitude which made for difficult relations with the electricity generation agency, particularly after it discovered he had set up an independent committee to review its operations. The head of the HEC, Russell Ashton, complained directly to Lowe about the poor relationship between Lohrey and the HEC. So, in mid 1978 Lowe sacked Lohrey from his hydro and forestry responsibilities and his ministerial responsibilities became restricted to the environment.

Lohrey brought to his administration of the NPWS the same independence of mind he demonstrated in his relationship with the HEC and was the best minister the NPWS ever had. He encouraged it to develop a formal proposal for a national park that would protect the lower Gordon and Franklin rivers. He knew all too well what a direct threat such a proposal would pose to the HEC. The NPWS developed the Gordon–Franklin national park idea with enthusiasm, arguing it was important for the state to have a national park that protected an entire water catchment.[20] The

NPWS even took the unprecedented step of promoting its national park proposal in the media. This was a daring move. Lohrey supported the action, arguing the HEC had used public funds to advance its case for the Franklin scheme and 'if it's good enough for the HEC, it's good enough for the National Parks and Wildlife Service'.[21] Lohrey also took part in the Franklin blockade. The NPWS's proposal marked the first occasion since the 1940s that the state's reserve management agency had stood up to major development interests. It was a refreshing return to form and an important addition to democratic decision-making in the state. Unfortunately, the NPWS's independence was short-lived and the Franklin dispute ultimately ended up being the high point of its autonomy.

The NPWS's Gordon–Franklin national park proposal was considered by Lowe's cabinet in its July 1980 deliberations on the HEC's proposal for the Franklin and lower Gordon rivers. Cabinet ultimately opted to build a compromise dam on another river that flowed into the Gordon—the Olga River—and to save the Franklin in a new Wild Rivers national park along the lines proposed by the NPWS. The Franklin–Gordon Wild Rivers National Park was pivotal to saving the Franklin River; its proclamation made the development of the Franklin River much more difficult. Aware of the strategic significance of creating the new national park, Doug Lowe said he 'indicated that the government would not proclaim the Wild Rivers National Park until such time as its legislation [authorising the compromise scheme on the Olga River] was passed by parliament'.[22] As noted in Chapter 3, by early 1981 it was obvious that the upper house of Tasmania's parliament would not pass the Olga scheme legislation. Lowe then took the crucial step of withdrawing the Olga legislation and proceeding with the proclamation of the Franklin–Gordon Wild Rivers National Park regardless of the upper house's intransigence.

On 13 May 1981, the 195 200-hectare Wild Rivers National Park was proclaimed along with a 39 000 hectare extension to the Southwest National Park. The new national park took in the former Frenchmans Cap National Park along with reserves that had earlier been created along the Gordon River and the Lyell Highway. Together with the Southwest and Cradle Mountain–Lake St Clair national parks, the Wild Rivers National Park was part of a continuous stretch of national parks that extended from the south coast through to Cradle Mountain.

Experience with past revocations from national parks had made it clear they were not sacrosanct and that it was possible a future government might revoke the new Franklin–Gordon Wild Rivers National Park. But Lowe decided to nominate the new national park for World Heritage listing, which he did in July 1981, enhancing its protection. Even though Lowe was toppled as premier in November 1981, his World Heritage nomination survived.

After Robin Gray was elected the new Liberal Premier of Tasmania in May 1982, pressure was placed on the Fraser government to withdraw the World Heritage nomination. But the federal government stood firm and a meeting of the World Heritage Committee in Paris in December 1982 considered it. In a desperate, last-ditch attempt to thwart the nomination, Gray sent his deputy premier to the meeting but his mission was fruitless. On 14 December 1982 the three national parks that made up the nomination were formally listed as the Western Tasmania Wilderness World Heritage Area, joining hundreds of other famous cultural and natural sites such as the pyramids of Egypt, the Grand Canyon, the Great Wall of China and Stonehenge. Lowe's crucial move had paid off.

What remained unresolved, however, was whether the listing necessarily gave the federal government the power to stop the Franklin dam. In April 1983 the new federal Labor government of Bob Hawke introduced legislation that gave effect to the listing, which passed through parliament the following month. After being contested in the High Court on 1 July 1983 the court ruled that the legislation was valid and that the Franklin could be saved. It was the finest hour yet for the NPWS, which felt more than vindicated about its criticism of the HEC's plans and its development of the Wild Rivers national park proposal. The victory largely came about because of campaigning by the environment movement but was made possible by Doug Lowe and Andrew Lohrey.

Post-Franklin restructuring of the National Parks and Wildlife Service

After the Franklin victory the river was safe but the NPWS was not. The pro-dam government of Robin Gray had been elected in 1982 on a platform that included removing the autonomy of the NPWS. It was well aware of the independent stance the service had taken during the Franklin dispute. The Gray government did not move immediately to muzzle the NPWS, but in 1986 it announced it would amalgamate the service with the Department of Lands.[23] In May 1987, the NPWS became part of a new Department of Lands, Parks and Wildlife. The service was placed within a vast, conservative department and lost its ability to give independent advice on development decisions with conservation implications. The merger marked a return to subservience of the state's reserve management agency, as had occurred with the Scenery Preservation Board in the 1950s and 1960s. The Director of the Tasmanian Conservation Trust, Phillip Hoysted, predicted that the amalgamation would 'save no money and lead to the watering down of the NPWS's role and integrity as well as the legislation which protected parks and reserves'.[24]

The amalgamation of the NPWS with the Lands Department proved to be the first of many changes to the service, most of which were designed to silence it as a voice for conservation. However, when the Labor government of Michael Field replaced Robin Gray's government in 1989, it returned some of the NPWS's autonomy by creating a separate Department of Parks, Wildlife and

Heritage. This was a condition of the Accord, the agreement of government signed with the Greens (see appendix 3).

But the new independence was short-lived. After the Field government lost office to the Liberal Party government of Ray Groom in 1992, yet another amalgamation took place. The Department of Environment and Planning was merged with the Department of Parks, Wildlife and Heritage in 1993 to create the Department of Environment and Land Management, in which the Parks and Wildlife Service (as it became known) was just a division.[25] When a new Labor government under Jim Bacon was elected in 1998, yet another merger happened in which the Department of Environment and Land Management was amalgamated with the Department of Primary Industries and Fisheries to form a new mega-department of Primary Industries, Water and the Environment.[26] The Parks and Wildlife Service was now lost within a huge bureaucracy with many layers of management between it and its minister. The days when minister Lohrey had worked closely with the independent NPWS were but a distant memory.

The final restructuring and dismembering of the Parks and Wildlife Service began in 2000 when the reserves management section of the Parks and Wildlife Service was given a separate divisional identity to the wildlife and heritage conservation functions. There had always been a symbiotic relationship between the service's conservation functions—concerned with fauna, flora, earth science and cultural heritage management—and the on-the-ground management of the state's reserves, but that link was severed by the change that came about largely as a result of pressure from the tourism industry. The tourism industry's representative body thought the separation would result in 'better decisions' being made about tourism development in the state's reserves.[27] The Tasmanian Greens condemned the change, calling it a 'theme park' approach to the management of national parks that could divert funding away from conservation functions.[28]

The separation was made complete in 2002 when the Bacon government removed the Parks and Wildlife division from the Department of Primary Industries, Water and the Environment, placing it within the Department of Tourism, Parks, Heritage and the Arts.[29] The separation was accompanied by the passing of two new acts in 2002 that replaced the 1970 *National Parks and Wildlife Act*. They were the *National Parks and Reserves Management Act* and the *Nature Conservation Act*. There was no pretence, any more, that national park management was connected with conservation. The government left little doubt that it viewed the management of national parks as part of the state's tourism industry. In many ways it was a fulfilment of Eric Reece's philosophy of making national parks into yet another resource industry as expressed in his 1969 election proposal to merge the Scenery Preservation Board with the office of the Director-General of Tourism.

The Tasmanian National Parks Association, an independent lobby group dedicated to improved national park management, opposed the 2002 move, arguing that an independent National Parks and Wildlife Service needed to be re-established that combined both conservation and park management functions. The organisation said it should be directed by someone with a 'long history of passionate involvement with nature conservation'.[30] Today, the Tasmanian government is deaf to any suggestion that such an independent conservation and reserve management body should exist. Within the state government there has been a change of culture that now largely accepts that tourism pressures are here to stay. At best, the state government these days is about adapting to tourism pressure rather than resisting it.

The 1989 expansion to Tasmania's World Heritage Area

The Parks and Wildlife Service was not directly involved in the doubling of Tasmania's World Heritage Area in 1989. The main influences were the Helsham Inquiry into the state's National Estate forests and negotiations between the Green Independents and the minority Field government. But the expansion had enormous implications for the service. In terms of the land it managed, the three national parks that made up the original Western Tasmania Wilderness World Heritage Area—the Cradle Mountain–Lake St Clair, Franklin–Gordon Wild Rivers and Southwest national parks—were joined by the Hartz Mountains and Walls of Jerusalem national parks in making up most of the enlarged World Heritage Area (now called the Tasmanian Wilderness World Heritage Area). All five national parks significantly increased in size—their combined area grew from 787 200 to 1 265 026 hectares. The Franklin–Gordon Wild Rivers, Southwest and Walls of Jerusalem national parks experienced the greatest increases. Conspicuous by their absence, however, were additions that could have included the forests of the middle reaches of the Huon, Picton and Weld rivers; the Tyndall Range and the area south of Macquarie Harbour; and the forests of the Great Western Tiers. All were excluded, like so many potential national park areas in Tasmania, because their apparently all-important development potential was prioritised over conservation.

The expansion of the World Heritage Area also made a big difference to the staffing and funding of the Parks and Wildlife Service. The service received significant extra funding from the federal government to pay for the management of the expanded World Heritage Area. This resulted in a large staff increase—throughout 1990 and 1991 the service was putting on new staff nearly every week. The three years between 1989 and 1992—when the service was both an independent department and was expanding its resources—were its finest hours since the 1970s but the freedom and growth did not last. As well as suffering the ignominy of being merged into larger departments, from 1994 the service suffered cuts in federal government funding of the Tasmanian World Heritage Area.

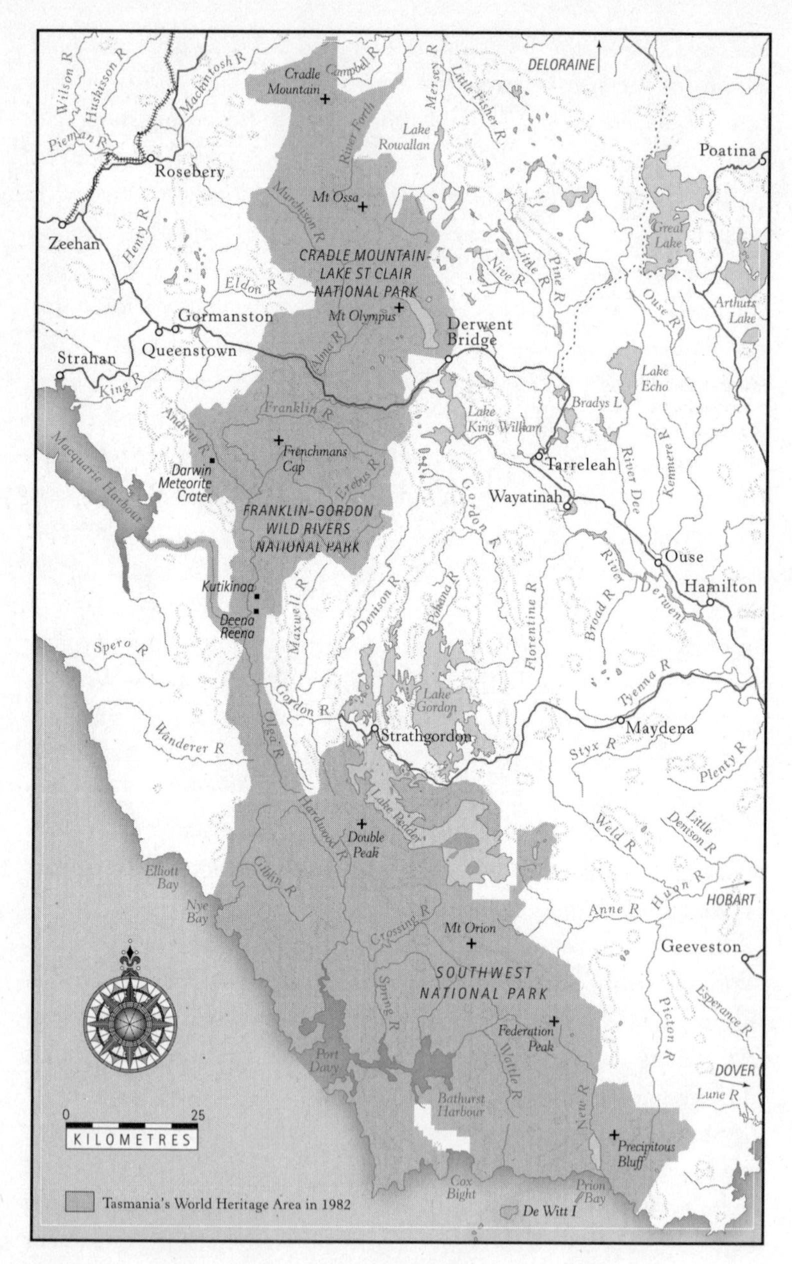

Figure 10.1 Tasmania's World Heritage Area in 1982.

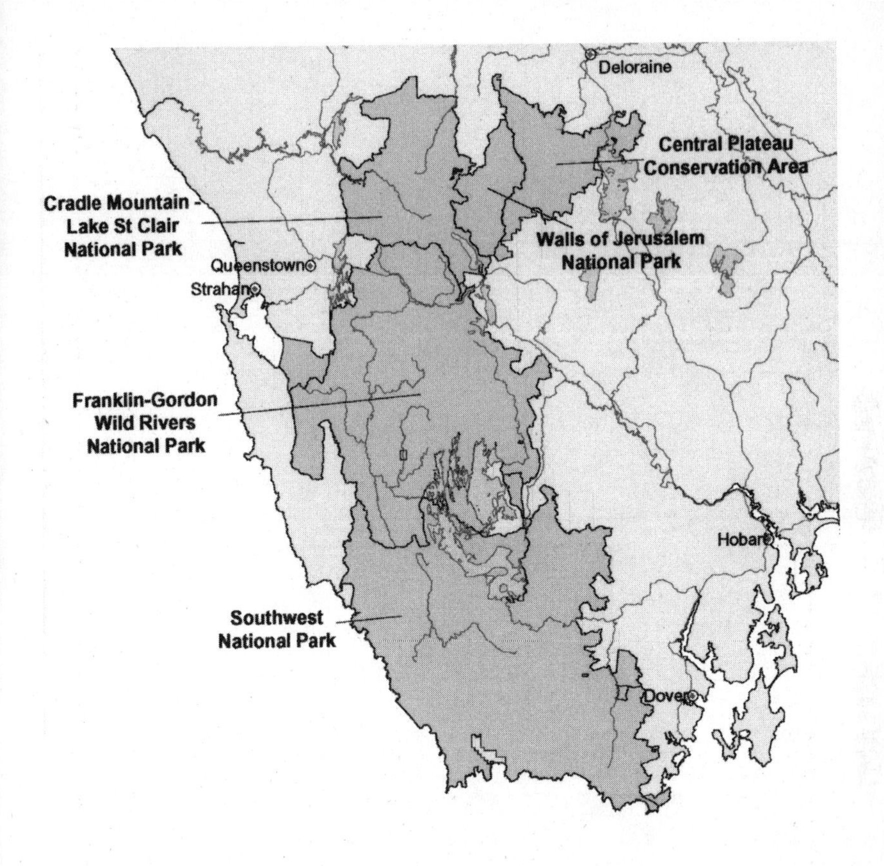

Figure 10.2 Tasmania's World Heritage Area after its expansion in 1989. (Source: Parks and Wildlife Service, Tasmania)

Another big change for the Parks and Wildlife Service came in the late 1980s with the creation of the Douglas–Apsley National Park in a region that embraced the last major unlogged catchment of forest in eastern Tasmania. The park's creation, a condition of the Accord between the Green Independents and the Labor Party struck in May 1989, was a victory against significant development interests that had long had designs on the area. The Shell petroleum company had coal exploration leases that covered nearly 40 per cent of the region while the woodchip company Tasmanian Pulp and

Forest Holdings (TP&FH) held a forest concession. TP&FH nearly started logging the Douglas–Apsley in 1987 but, fortunately, was thwarted by a moratorium declared to allow investigation of the area's forests.[31] The government of Robin Gray rejected the idea of creating a Douglas–Apsley national park but after the Greens and Labor Party came to power in late 1989, a 16 080-hectare national park was finally created in the area.[32]

Not only was the Douglas–Apsley's environment unique, so too was the politics behind its creation. The park was the first to be created in Tasmania with enthusiastic support from nearby residents and businesses. By the late 1980s national parks had turned, in many Tasmanians' eyes, from being liabilities to assets. Suddenly, Tasmanians could see the tourism and aesthetic benefit of having a national park created near their community, although there were still powerful resource and traditional use interests steadfastly opposed to the creation of new national parks. The new attitude proved both a blessing and a blight for the Parks and Wildlife Service. In the mid 1990s it resulted in the Liberal government of Premier Ray Groom proposing two new national parks—one along the south coast of Bruny Island, south-east of Hobart, and another around the limestone caves of Mole Creek in the north. But the new-found enthusiasm for national parks also put these areas, and other parks, in danger of being exposed to unsustainable tourism pressure.

As early as 1980 the NPWS had recommended that the coastal fringe of the southern part of Bruny Island—which includes extensive beaches, large dolerite rock cliffs, pristine coastal forest and unique heathland environments—be made a national park. Robin Gray's government considered the idea but baulked at a proposal that some private land be purchased to be included in the park. It ended up doing nothing.[33] The new national park proposal, however, had considerable backing from residents and businesses on Bruny Island, as well as from the Greens and the Labor Party. So the idea was revisited in the mid 1990s when the Forestry Commission indicated it was prepared to revoke 800 hectares of state forest for inclusion in the park. There was no major opposition to the new park, so in 1997 the 5149-hectare South Bruny Island National Park was proclaimed, although it was largely made up of existing reserves.

The Mole Creek Karst National Park was created the year before the South Bruny National Park, in 1996, to protect some of the approximately 200 caves situated around the town of Mole Creek. As early as 1899 the significance of these caves had been recognised when they became some of the first natural environment reserves created under the *Wastelands Act*. As they had for the South Bruny National Park, local residents and businesses supported the park, recognising its tourism potential. The park itself was not contiguous but was a collection of nine separate reserves covering 1345 hectares, most of which had been reserves before being part of the national park.

National park changes from the 1997 Regional Forest Agreement

The Parks and Wildlife Service did not play a significant part in the 1997 Regional Forest Agreement (RFA) but its signing, like the 1989 expansion of the World Heritage Area, led to significant expansion of the land managed by the service. The RFA fell a long way short of adequate reservation of the unprotected forests of Tasmania. The Tasmanian Conservation Trust calculated that it should have resulted in about 1.2 million additional hectares of land being reserved,[34] but instead just 396 000 new hectares were reserved, of which 58 259 hectares were added to the state's national parks and other state reserves. The Southwest, Franklin–Gordon Wild Rivers, Cradle Mountain–Lake St Clair, Freycinet and Mount William national parks had small additions made to them while two new national parks were created—the

Savage River and Tasman national parks—which together reserved an additional 28 730 hectares.

The 17 980-hectare Savage River National Park was a small victory for the conservation movement, which had long campaigned for meaningful protection of the rainforests of the Tarkine area in the north-west—the largest continuous area of temperate rainforest in Australia. But it was only a small victory. The new national park protected less than 10 per cent of the more than 200 000 hectares of myrtle, celery top pine, sassafras, leatherwood and blackwood rainforest that exist in the area. Within the national park a significant proportion of the reserved land was ridgetop environments that had no forest on them.[35] It was nonetheless a step forward. The idea of a national park in the region had been around since the early 1970s when activist Peter Sims put a proposal to the Scenery Preservation Board for a Norfolk Range national park. This was nearly adopted by the Liberal state government of Angus Bethune before it fell to the Labor government of Eric Reece (see Chapter 4).

The Savage River National Park was unique for Tasmania because it was the first and only national park in the state that had no reliable road access to it. It was created wholly and solely for conservation purposes. In 2005, 73 500 hectares of new land was added to the protected area of the Tarkine as part of the Lennon–Howard Community Forest Agreement (see Chapter 6), but none of the newly protected land was added to the Savage River National Park and only 31 000 hectares was made into new reserves of any kind.

Like the Savage River National Park, the 10 750-hectare Tasman National Park, created around the southern and eastern coastal strip of the Tasman Peninsula, east of Hobart, resulted in protection of little additional forest to that which had already been reserved. The Tasman Peninsula area includes stunningly beautiful shoreline dominated by some of the tallest sea cliffs in Australia. It also embraces spectacular beaches, unique coastal heath and several

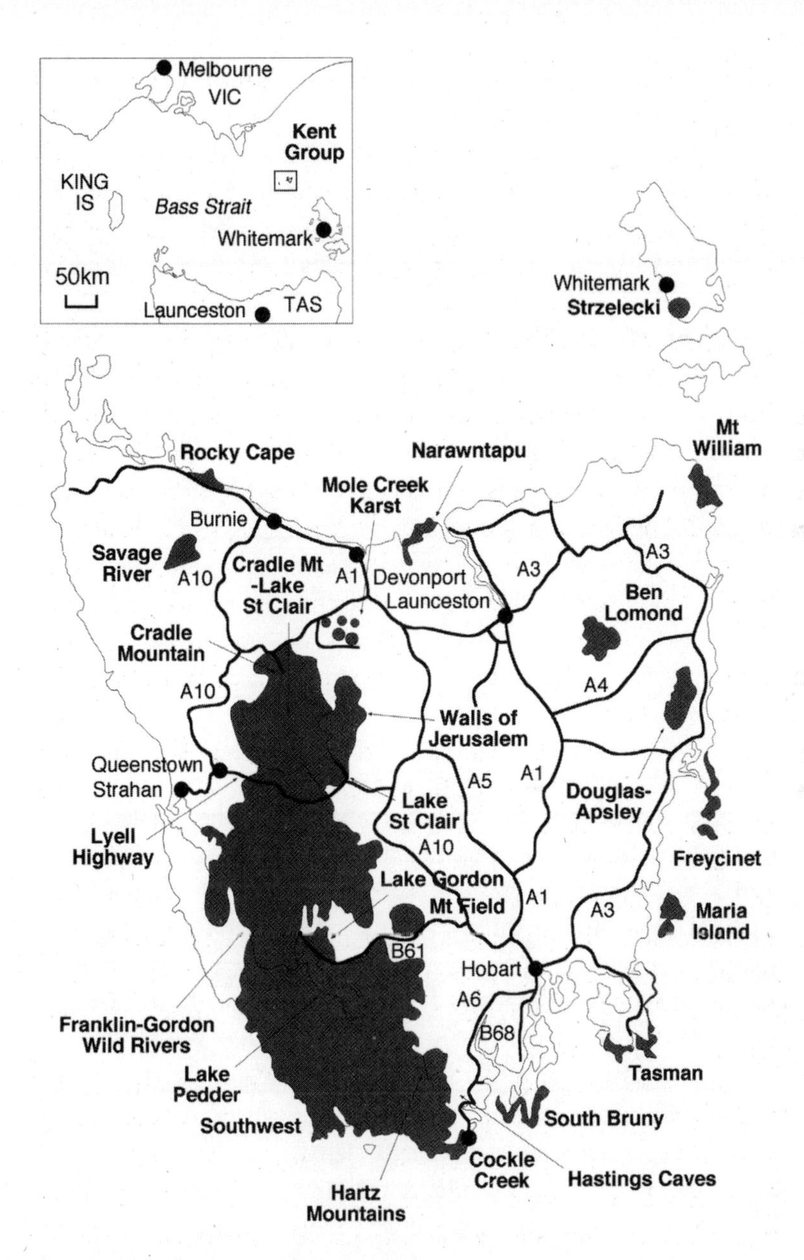

Figure 10.3 Tasmania's current national parks. (Source: *Tasmania's National Parks: A Visitor's Guide*, published by Greg Buckman, 2002)

offshore islands that are home to seals and many bird species. Since the early 1990s local conservation activists, headed by one-time Franklin River activist Peter Storey, had lobbied for a national park in the area. Prior to the declaration of the national park a large part of its land was reserved in small, separate reserves. In 1995 the state Land Use Commission, an autonomous government body that managed state land tenure, recommended the reserves be combined into a national park. The state government set up an interdepartmental committee to examine the idea and a majority backed the national park concept.[36]

Although the development of the national park was independent of the RFA, it eventually came to be tied to it. The park did not result in any previously unreserved land being protected but did result in some land previously managed by the state forestry service—Forestry Tasmania—being transferred to the Parks and Wildlife Service.[37]

Tourism pressure on Tasmania's national parks

However flawed their creation might have been, the establishment of the new Mole Creek Karst, South Bruny Island, Savage River and Tasman national parks was testament to the popularity of national parks in Tasmania. Tasmania's wilderness, particularly in its national parks, had become famous. Although the fights over Lake Pedder in the 1970s, the Franklin River in the 1980s and the state's forests in the 1990s were fraught, they seared into the consciousness of most Australians an awareness of Tasmania's unique wild areas. Tasmania came to stand for natural beauty—the state's vehicle numberplates even declared it to be 'your natural state'. Advertisements for the state's tourism industry came to be based on sumptuous shots of its unique environment and the name 'Tasmania' became equated to the great outdoors and untamed beauty.

This new awareness was a potential source of support for the conservation movement. But there was an accompanying liability in the form of increased pressure from tourism. As more and more tourists visited Tasmania, demands began to mount to develop its national parks in ways that did not necessarily respect conservation priorities.

The first major push to develop the state's national parks came in 1985. The Liberal government of Robin Gray called for expressions of interest from private developers who might be interested in establishing a series of commercial walking huts along the popular Overland Track in Cradle Mountain–Lake St Clair National Park. There had never been commercial walking huts in Tasmania's national parks before and the move seemed like an attempt to effectively part-privatise the track. The huts would be for the exclusive use of one operator and would be situated well away from the track's public huts. The huts proved popular and were later expanded, but they set a disturbing precedent, sending out a signal that the state's national parks were 'open for business'. Pressure grew for private access rights to the state's other national parks. Although commercial huts have not been developed, a number of operators have been given the right to have exclusive 'standing camps' in national parks such as Freycinet and Maria Island. These allow semi-fixed bush accommodation to be erected for most of the year for sole use by company clients. Although not as intrusive as permanent commercial huts, the standing camps still grant exclusive business rights to part of a national park. They dilute the notion that the parks belong to everyone. Commercial huts have been proposed for an upgraded track to be built in Tasman National Park.

By the late 1990s, the Tasmanian tourism industry was becoming well aware of the business potential of the state's national parks. In 1999 the industry's peak body, the Tasmanian Tourism Council, produced a report that was critical of the Parks and Wildlife Service, claiming the service had an 'obstructive' bureaucracy that made it difficult for ecotourism operators to expand.[38]

Although the report was not intended for public release, the media reported on it and soon after the Labor government of Jim Bacon announced that the service would be 'stripped of its exclusive power to process wilderness developments'.[39]

Straight after that change, the government indicated it might allow tourist helicopters to land in national parks and began receiving submissions from interested operators. It then released a list of five potential sites for helicopter landing pads.[40] Landings would be very intrusive, particularly in the national parks that make up the state's World Heritage Area as helicopters make a lot of noise, adversely affecting the animals of the parks as well as compromising the wilderness experience for visitors hoping to escape the distractions of modern living. The 'chopper debate' created a substantial public backlash and out of 651 public submissions received by the government on the proposal, all but twelve opposed it. Even many small ecotourism operators were against it, arguing their businesses would suffer if their clients could no longer get peace and solitude from their national park visits.[41] The reaction was sufficient to make the government retreat. The issue made the conservation movement aware of the threat that tourism could pose to the state's national parks. It prompted the creation of a new body dedicated to the defence of proper national park management, the Tasmanian National Parks Association.

The tourism threat did not end with the helicopter debate. A year later, in 2001, the state government announced it had approved the development of a lodge and some cabins within the Southwest National Park at Cockle Creek on the south-east coast. Like the commercial huts on the Overland Track, this intrusion reinforced the notion that the state's national parks could effectively be part-privatised. It amounted, in many ways, to a revocation of part of the national park.

Lodges and temporary or permanent commercial huts are the favoured type of development of the tourism industry in Tasmania's national parks. In 2005 approval was given for new private lodges to be built on the shore of Lake St Clair, within the Cradle Mountain–Lake St Clair National Park. Another lodge developer is likely to be given approval to redevelop old Hydro Electric Commission buildings as a luxury lodge further around the lake at Pumphouse Point. Permission was also given to a private developer to redevelop some of the historic structures as tourist accommodation on Maria Island. Such is the apparent power of tourism in Tasmania these days, that a new national park was created in 2001 in the Kent Group (a small group of islands north-west of Flinders Island) largely at the behest of a potential developer. The march of tourism pressure seems unstoppable. Many areas have been saved only to be exposed to new development pressures brought by visitors who want boutique access to the state's wild areas.

There is no evidence that the new tourism pressure is supported by the Tasmanian public. A poll conducted by the Launceston *Examiner* found 73 per cent of respondents opposed development within national parks.[42] Tourism development is the greatest threat confronting the state's national parks, while sustainable tourism management is *the* greatest challenge. Developers do not need access to prime sites within the state's national parks to make their ecotourism businesses work. At Cradle Mountain there are many private lodges situated outside park boundaries that do very well. But the government is forever lobbied by developers who want a site that no other operator can have. If badly managed, tourism will be the death of many of the special qualities of Tasmania's national parks because it can lead to the destruction of their best qualities.

The inadequacy of Tasmania's reserves

By 2005 Tasmania had nineteen national parks that covered 1 431 305 hectares, or 21 per cent of the state. It also had another 570 reserves, most administered by the Parks and Wildlife Service, that covered an additional 1 174 955 hectares or 17 per cent of the state.[43] Despite this apparently significant area of preservation, the state's reserve system is patchy and inadequate.

Tasmania is sometimes divided into nine 'bioregions', according to ecosystem type, and amongst those bioregions reservation is very uneven. In the West bioregion 83 per cent of all land is reserved; in the Central Highlands bioregion 56 per cent is reserved; and in the Southern Ranges bioregion 44 per cent is reserved. But that is where the good news ends. In all of the other six bioregions, less than 20 per cent is reserved—in the Northern Midlands bioregion just 2.6 per cent is reserved.[44] Tasmania does have an extensive network of reserves but it is heavily biased towards areas that industry is not interested in, generally in the western half of the state. Huge expanses of buttongrass and alpine woodlands are reserved but precious few wetlands, estuaries and marine habitats are, for instance.[45] Many of Tasmania's forest species—which the RFA was meant to protect—are also under-reserved. Nearly all of the state is now carved up and it is going to be difficult to achieve more reservation without wholesale changes in tenure, such as a significant amount of state forest being transferred into national park. But such change needs to take place before the state's reserves can become truly representative.

Tasmania is left with a patchy reserve system that is the product of many compromises designed to appease development interests, and is under enormous pressure from the tourism industry. Although Tasmanians take pride in their state's natural areas, the state is a long way from managing them in a truly sustainable way. In short, Tasmania has a compromised reserve system out of which too many people want to make money. The ethos that says land should only be reserved in a national park if it has no other economic uses, which dominated the thinking of the US Congress when it created the world's first national park more than 130 years ago, is still dominant. Tasmania has not moved on. The state has yet to substantiate its claim to being 'your natural state'.

11

Conclusion

Tasmania's wilderness battles have had it all: violence, bribery, arson, lawsuits, anti-democratic legislation and lots of dodgy deals. Despite the dramas and enormous resources marshalled against it, the Tasmanian conservation movement has chalked up some major gains in the clashes; most notably the stopping of the Franklin dam and the creation, and enlargement, of the Tasmanian Wilderness World Heritage Area. Other big wins have included the saving of the Tarkine rainforests, the north Styx forests, the elevated Great Western Tiers forests, the forests of the upper Huon, Picton and Weld rivers, the Recherche Bay forests, the Douglas–Apsley forests, Precipitous Bluff, and the Beech Creek and Counsel River forests. The movement has also stopped the proposed Wesley Vale and Huon Forest Products pulp/woodchip mills, has seen the end of the conversion of native forest into plantations, and has gained standing in environmental court cases.

But the movement has also had some major setbacks: Lake Pedder has been flooded (at least for now); the upper Pieman and Murchison rivers have been flooded as has the Forth Falls; the natural flow of the Cataract Gorge has been taken away; the forests of the middle Picton, Huon and Weld rivers are being logged as are the south Styx forests and most of the state's east coast forests; almost 100 000 hectares of native forest have been converted to plantations; and the King and Queen rivers have been polluted beyond easy remediation. Crucially, the Hartz Mountains, Mount Field, Lake Pedder, Freycinet and Cradle Mountain–Lake St Clair

National Parks have been revoked, at least in part, making Tasmania the worst state for national park revocations in the country. Tourist lodges, or semi-permanent standing camps, have also been set up within the boundaries of the Southwest, Mount William, Freycinet, Maria Island and Cradle Mountain–Lake St Clair national parks. Worst of all, Gunns' Tamar Valley pulpmill has been given permission to go ahead.

The net effect of the conservation movement's wins and losses has been that within the last century the amount of public land set aside in reserves has climbed from less than 1 per cent of the state in 1910 to 39 per cent today, but all the undeveloped public land outside that 39 per cent is more threatened than ever. Tasmania has basically put a fence around some of its wilderness areas—mainly the uneconomic bits—and has set about destroying the rest.

The biggest loss, by far, has been trust. It has hardly ever existed in Tasmania's wilderness battles despite its brief life during the 1990 Salamanca Agreement forestry process. The lack of trust between protagonists has been an indelible feature of all the state's environmental battles since Lake Pedder and the Franklin River campaign.

Albert Einstein once said, 'you cannot solve a problem with the same consciousness that made it'. Tasmania's problem is that too much of its thinking clings to a resource exploitation philosophy rooted in the 1950s. There has not been a change of consciousness so the state keeps returning to the same wilderness development problems time and time again. Several states have shifted their forestry industries from native forest to plantation forests but Tasmania keeps clinging to native forest logging. Most states jealously guard their tourist image and keep their national parks intact, but Tasmania seems all too prepared to sacrifice both to more development. The Tasmanian government pays lip service to the state's clean, green image but does little to justify it. Photographer Matthew Newton says the state has a 'seemingly schizophrenic government [that] promotes the place as the last great wilderness destination on the planet, whilst ripping the guts out of its majestic old-growth forests'.[1]

Essentially, Tasmania is still pursuing Premier Albert Ogilvie's three-pronged Depression-era strategy of boosting power consumption, expanding the forestry industry and increasing access to unique wilderness destinations. Ogilvie's time was long ago: wilderness is scarcer and more important now and Tasmania needs to shift to reflect the changing times. It cannot stay forever locked in the early twentieth century. It has to move on.

Tasmania's lack of forward-thinking makes for deep cynicism. Author Richard Flanagan says that government in Tasmania has degenerated into 'a culture of secrecy, shared interest and intimidation that seems to firmly bind the powerful'.[2] He laments: 'it is little wonder that many Tasmanians now worry that the woodchippers' greed destroys not only their natural heritage, but distorts their parliament, deforms their polity and poisons their society'.[3] Evidence of this poison is all too easy to find in the state's wilderness battles.

The common cause of these battles has been an all-too-cosy relationship between Tasmanian governments and resource developers. Generally, there has been an obsequious attitude by Tasmanian governments to resource interests that is as undiminished today as it was in the late nineteenth century when the Mount Lyell mine was established. When commenting on the close relationship between Gunns and the Tasmanian government in 2007, Marcus Priest from the *Australian Financial Review* wrote: 'it is symptomatic of the way business and politics are conducted in the state, where there are only ever two degrees of separation at most'.[4]

All too often the lack of adequate separation has resulted in secret deals that have undermined Tasmanian democracy in the way Flanagan suggests: in the Florentine Valley clash in the 1940s, in the Lake Pedder fight in the 1960s, in the flooding of the Pieman River in the 1970s and in the 2007 battle over Gunns' proposed pulpmill, the Tasmanian government even introduced special legislation that denied democratic appeal rights. The *Scenery Preservation Act* was twice watered down to make development in national parks easier. Draconian trespass laws were passed to prevent the environment movement protesting on the Gordon River or in threatened forests. Gunns has tried to bring punitive lawsuits against forest activists. The Parks and Wildlife Service has been muzzled. The list goes on. There seems no limit to the lengths government and resource industries will go to thwart opposition to the development of Tasmania's forests, rivers, minerals and national parks.

But there is an emerging vision of a new and sustainable consciousness in Tasmania that could lead to more enlightened decision-making if only the state government would see it. It is a consciousness based mainly around small- to medium-sized businesses that trade on Tasmania's clean, green image. It is made up of ecotourism operators who respect the sanctity of the island's national parks, vineyards that trade on Tasmania's pristine image, consultants who work from the state because of its unique lifestyle, designers who draw their inspiration from the island's natural beauty, farmers who benefit from the state's unpolluted soils and manufacturers that have carved sustainable niches in areas like confectionery, lotions and beverages. They are living, breathing examples of the direction Tasmania could take if only the state had the courage to embrace it. Essentially, the state needs to replace its twentieth-century, factory-led development mindset with a twenty-first century mindset that focuses on the development of small business and services. Member of the Tasmanian Roundtable for Sustainable Industries, John Dingamanse, encapsulated the new direction when he declared in 2007 in relation to the Gunns' pulpmill debate: 'my personal belief is that rather than the 60s and 70s culture of heavy industrialisation, Tasmania's future lies in smaller tourism projects that provide greater flow-on benefits'.[5]

Tasmania needs to break with its past; to look at its unique wilderness areas anew and to step, wholeheartedly, into the sustainable future that many people and businesses are already creating for the state. Tasmania needs to reach into its soul and discover a new meaning of wilderness: a meaning based on inspiration and sustainability; a meaning connected with the brighter parts of its consciousness instead of ones based on short-term thinking. In 1998 the Greens told Tasmanians: 'If you love this island . . . you'll have to vote for it'. This means voting for an enlightened view of wilderness. It means creating a new Tasmania.

APPENDICES

Appendix 1
Key dates in Tasmania's wilderness battles

Nineteenth century

1824 Peter Degraves establishes Tasmania's first permanent sawmill

1858 New *Wastelands Act* allows for the reservation of public land for the first time

1864 George Perkins warns about the environmental impact of forestry, agricultural land clearing and industrial pollution in his book *Man and Nature*

1871 James 'Philosopher' Smith discovers a major tin deposit at Mount Bischoff, starting a mining boom in Tasmania

1872 The world's first national park, Yellowstone National Park in the United States, is created

1879 Australia's first national park, Royal National Park south of Sydney, is created

1881 The Tasmanian *Wastelands Act* is amended to allow for the charging of forestry royalties for the first time

1885 Russell Falls, as well as some Central Plateau lakes, become the first significant natural area reserves created in Tasmania

1891 Tin mining begins at Cox Bight in south-west Tasmania

1895 Tasmania's first hydro electric power station, at Duck Reach on the South Esk River, commences operation

1896 Smelting begins at the Mount Lyell copper mine, leading to the devastation of the surrounding environment

1898 The state Conservator of Forests, George Perrin, warns about the overcutting of Tasmania's forests

Twentieth century

1903 The new *Crown Lands Act* supersedes the *Wastelands Act*. After amendment in 1911 it allows for easier reservation of public land

1915 A new Tasmanian Labor government passes the *Scenery Preservation Act,* the most progressive nature preservation legislation in Australia to date

1916 Waddamana, the Hydro Electric Department's first power station, commences operation

1916 The first hydro power contract is signed with a bulk user, Amalgamated Zinc

1916 Tasmania's first reserves, that are eventually made into national parks, are created at Mount Field and Freycinet

1921 The *Scenery Preservation Act* is watered down to allow for resource development within reserves

1922 A new 'floatation' smelting process at the Mount Lyell mine begins nearly a century of tailings dumping into the Queen and King rivers

1922 The Cradle Mountain–Lake St Clair National Park is created

1925 Significant osmiridium mining begins at Adamsfield

1929 The Hydro Electric Department is made into a commission, giving it considerable independence

1935 Significant gold mining begins on the Jane River

1935 Granite mining begins within Freycinet National Park

1937 APPM begins operating its Burnie pulp mill, Tasmania's first pulp or woodchip mill

1938	Minerals are discovered in the Pelion area of Cradle Mountain–Lake St Clair National Park, leading to the revocation of 1200 hectares of the park
1939	Hartz Mountains National Park is created
1940	The Hydro Electric Commission oversees the raising of Lake St Clair by 3 metres
1941	ANM begins operating its new Boyer pulp and paper mill near New Norfolk
1941	Frenchmans Cap National Park is created
1941	Charles King takes over tin mining lease at Melaleuca
1943	Prime Minister John Curtin announces that Australia's first aluminium smelter will be built in Tasmania
1943	1214 hectares are revoked from Hartz Mountains National Park for forestry
1946	Tasmania's first significant conservation group, the Tasmanian Fauna and Flora Conservation Committee, is formed
1946	A number of Tasmania's major reserves are renamed as national parks to increase their appeal
1947	Ben Lomond National Park is created
1950	The *National Park and Florentine Valley Act* is passed, excising 1523 hectares of tall forest from Mount Field National Park
1953	The HEC puts the first flow metres on the Gordon River in preparation for the Lake Pedder hydro scheme
1955	The new Trevallyn hydro scheme significantly reduces water flow through Launceston's Cataract Gorge
1955	Lake Pedder National Park is created
1957	Deny King completes an airstrip at Melaleuca
1958	The HEC indicates interest in developing new hydro schemes on the Mersey, Forth, Gordon, Franklin, Pieman and King rivers
1962	Premier Reece announces that the Gordon River will be investigated as the potential site of three new hydro schemes

1962 The South West Committee conservation group is formed
1962 The HEC builds a jeep track into the Serpentine River near Lake Pedder
1965 Premier Reece admits there will be 'some modification' to Lake Pedder
1965 Entrepreneur Roy Hudson and the Pickards Mather company announce they intend to build a major mine at Savage River
1967 The HEC tables its official report on the Lake Pedder–Lake Gordon hydro scheme
1967 The Save Lake Pedder National Park conservation group is formed
1967 Legislation authorising the Lake Pedder–Lake Gordon hydro scheme is passed
1967 Rocky Cape National Park is created
1967 Strzelecki National Park is created
1967 The Tasmanian upper house inquiry into the Lake Pedder hydro scheme recommends the scrapping of the Scenery Preservation Board
1968 Permission is given to Tasmanian Pulp and Forest Holdings to establish a woodchip mill at Triabunna, the state's first woodchip mill
1968 Mining begins at Savage River on the southern edge of the Tarkine wilderness
1968 Southwest National Park is created
1968 The Tasmanian Conservation Trust conservation organisation is formed
1970 The HEC begins investigating the Franklin River for a future hydro scheme
1970 Olegas Truchanas succeeds in getting 400 hectares of Huon pine reserved on the Denison River
1970 Tasmania's *National Parks and Wildlife Act* is passed, allowing for professional management of its reserves for the first time

1971 The National Parks and Wildlife Service begins operating at the same time as the Scenery Preservation Board folds

1971 The Lake Pedder Action Committee conservation group is formed after more than 1000 people visit the lake

1971 The United Tasmania Group is formed, the first Green party in the world

1972 Olegas Truchanas drowns on the Gordon River

1972 Brenda Hean and Max Price die in mysterious circumstances while flying a light plane to Canberra to protest the flooding of Lake Pedder

1972 The Tasmanian Mining Warden upholds an objection from the Tasmanian Conservation Trust, and other groups, to the granting of a licence to Mineral Holdings for mining at Precipitous Bluff

1972 North-west activists led by Peter Sims lose the campaign to get the Norfolk Range made a national park

1972 Maria Island National Park is created

1972 The army is forced to back down over plans to turn Cockle Creek into an artillery firing range

1972 A federal government inquiry recommends a moratorium on the flooding of Lake Pedder; Premier Reece rejects the idea

1973 The Tasmanian Supreme Court finds the Tasmanian Conservation Trust had no legal standing with which to object to the Precipitous Bluff mining licence

1973 Mount William National Park is created

1973 The Whitlam government signs the World Heritage Convention

1974 The South West Action Committee conservation group, precursor to the Tasmanian Wilderness Society, is formed

1974 Tasmania's first campaign against woodchipping is mounted at Meander

1975 A draft new Southwest National Park management plan recommends excision of part of Hartz Mountains National Park in return for inclusion of Precipitous Bluff in the national park

1975 The full bench of the Tasmanian Supreme Court upholds an earlier decision that the Tasmanian Conservation Trust has no legal standing in the Precipitous Bluff mining case

1976 Bob Brown and Paul Smith make the first of several trips down the Franklin River

1976 The Tasmanian Wilderness Society (TWS) conservation group is formed

1976 2150 hectares is excised from Hartz Mountains National Park and given to Australian Paper Manufacturers in return for the reservation of Precipitous Bluff

1976 Asbestos Range National Park is created

1976 The Southwest National Park is doubled in area to about 400 000 hectares

1977 The High Court upholds the earlier Tasmanian Supreme Court decision that the Tasmanian Conservation Trust had no legal standing in the Precipitous Bluff mining case

1979 The Lake Pedder–Lake Gordon hydro scheme is opened

1979 The HEC announces details of the hydro scheme it has planned for the Franklin River

1979 The Lowe government backs the compromise Gordon-above-Olga hydro scheme

1979 Tasmania's upper house refuses to pass the Gordon-above-Olga hydro scheme legislation

1981 A Tasmanian referendum on the south-west hydro schemes yields a 45 per cent informal vote

1981 Walls of Jerusalem National Park is created

1981 Wild Rivers National Park is created

1981 Legislation authorising the Franklin River hydro scheme is passed (in June) after Robin Gray's Liberal Party wins government

1982 The 'Western Tasmania Wilderness National Parks' are inscribed onto the World Heritage list on the same day (in December) that the Fraser government says it will not intervene in the Franklin River issue

1982 The Franklin blockade starts (in December)

1983 The Labor Party, under Bob Hawke, wins federal government (in March) on a platform that includes saving the Franklin River

1983 The High Court (in July) rules in favour of the federal government's power to stop the Franklin Dam

1985 The Hawke federal government renews Tasmania's woodchip licences for fifteen years, giving the state forestry industry nearly all the forests it wants to log

1985 The Gray government allows private, commercial huts to be built on the Overland Track

1986 Bob Brown and TWS release a book of Lake Pedder photographs

1986 Major forest blockades are held in the Lemonthyme and Farmhouse Creek forests (in March); clashes occur between loggers and conservationists at both sites

1986 The Hawke federal government establishes the Helsham forests inquiry after Premier Gray authorises logging at Jackeys Marsh–Quamby Bluff

1987 North Broken Hill and Noranda announce plans to build a pulpmill at Wesley Vale that will use a highly polluting kraft chlorine pulping process

1987 Plans are announced for a new woodchip mill at Whale Point on the Huon River to be built by Huon Forest Products

1987 In the first of several changes, the National Parks and Wildlife Service is amalgamated with the Department of Lands

1988 The Helsham Inquiry finds (in May) that only 10 per cent of the forests it assessed are worthy of World Heritage listing but Commissioner Hitchcock publicly disagrees, as do most of the inquiry's consultants

1988 The Hawke federal government decides (in August) to list for World Heritage a major part of the forests considered by the Helsham Inquiry but the Gray state government gets a federal government undertaking that it will not conduct any more forest inquiries and will allow an increase in Tasmania's woodchip exports

1989 The proposed Wesley Vale pulpmill is stopped (in March) when Noranda pulls out of the project

1989 The Accord governing agreement between Green Independents and the Labor Party (signed in May) kills off the proposed Huon Forest Products Whale Point woodchip mill

1989 The Accord governing agreement establishes the Salamanca Agreement review of alternatives to logging Tasmania's National Estate forests

1989 Douglas–Apsley National Park is created

1989 Additions to Tasmania's World Heritage Area, which double its area, are agreed to by the Green Independents and the Field Labor government

1990 A Tasmanian state government inquiry rejects draining Lake Pedder

1990 The Salamanca Agreement process collapses (in September) as a result of intense pressure from the Field government

1992 The Field Labor government loses office (in February) after its relations with the Green Independents break down the year before as a result of its enthusiasm for forestry 'resource security' legislation

1992 The federal environment minister, Ros Kelly, announces that Benders Quarry will close the following year

1994 *Pedder 2000*, a conservation group dedicated to draining Lake Pedder, is formed

1994 The last hydro scheme in Tasmania is opened

1994 The federal resources minister, David Beddall, renews woodchip licences in most of 1311 sensitive forests around Australia, including most in Tasmania

1995 In a case brought by the Tasmanian Conservation Trust, the federal court rules (in January) that the federal government did not follow proper procedure when granting Gunns its first woodchip licence

1995 After log trucks ring Parliament House in Canberra, the Keating government grants woodchip licences for all but a few hundred sensitive forests around Australia and begins negotiations with states about new Regional Forest Agreements

1996 The 53-kilometre Tarkine Road—'The Road to Nowhere'—is opened

1996 Mole Creek Karst National Park is created

1997 The Tasmanian Regional Forest Agreement (RFA) is signed (in November). It is a disappointment to the conservation movement because it results in inadequate forest reservation and abolishes the state woodchip limit

1997 South Bruny National Park is created

1998 A major post-RFA blockade, attended by over 1000 protesters, is held at Mother Cummings Peak

1998 The electricity retailing and distribution arms of the HEC are separated into independent organisations

1999 Savage River National Park is created

1999 Tasman National Park is created

Twenty-first century

2000 Gunns buys the woodchip operations of Boral

2001 Gunns buys the woodchip operations of North Forest Products, making it a monopoly woodchip exporter in Tasmania

2001 Kent Group National Park is created

2002 The national parks division of the Parks and Wildlife Service is moved to a different department to its conservation division

2002 The *National Parks and Reserves Management Act* and the *Nature Conservation Act* replace the 1970 *National Parks and Wildlife Act*

2003 Gunns first indicates an interest in building a large pulpmill in northern Tasmania

2004 Gunns serves the 'Gunns 20' writ (in December) on various conservation groups and activists

2005 Gunns announces (in February) that its proposed pulpmill will use a chlorine process despite earlier assurances it would not

2005 The state/federal government Lennon–Howard Community Forests Agreement reserves much of the Tarkine and about half of the Styx forests but saves little else (in May)

2006 The Basslink electricity cable between Tasmania and Victoria begins operating, causing scouring of the Gordon River

2006 Bob Brown secures a deal with entrepreneur Dick Smith (in February) to buy threatened forests at Recherche Bay

2006 Bob Brown wins a federal court case (in December) about the impact of logging on threatened species in the Wielangta forests

2006 Gunns drops its legal action against five of the original 'Gunns 20' defendants after its first three writs are rejected by the Victorian Supreme Court (in December)

2007 Two Resource Planning and Development Commission (RPDC) panel members, who were assessing the Gunns proposed pulpmill, resign (in January)

2007 Gunns withdraws from the RPDC pulpmill assessment process (in March); the Lennon government then puts in place a lesser fast-track approval process

2007 Tasmania's parliament passes legislation giving state approval to Gunns' pulpmill (in August)

2007 The federal environment minister, Malcolm Turnbull, gives Commonwealth government approval to Gunns' pulpmill (in October)

2007 Bob Brown loses an appeal (in December) against his 2006 court win concerning logging in the Wielangla forests

2008 Conservation groups prepare for a possible blockade of the construction of Gunns' pulpmill

Appendix 2
The New Ethic

1974 manifesto of The United Tasmania Group, the world's first environment party

We, citizens of Tasmania and members of the United Tasmania Group,

United in a global movement for survival;

Concerned for the dignity of man and the value of his cultural heritage while rejecting any view of man which gives him the right to exploit all of nature;

Moved by the need for a new ethic which unites man with nature to prevent the collapse of the life support systems of the earth;

Rejecting all exclusive ideological and pragmatic views of society as partial and divisive;

Condemning the misuse of power for individual or group prominence based on aggression against man or nature;

Shunning the acquisition and display of individual wealth as an expression of greed for status of power;

While acknowledging that Tasmania is uniquely favoured with natural resources, climate, form and beauty;

Undertake to live our private and communal lives in such a way that we maintain Tasmania's form and beauty for our own enjoyment and for the enjoyment of our children through unlimited future generations;

Undertake to regulate our individual and communal needs for resources, both living and non-living, while preventing the wholesale extraction of our non-replenishable resources for the satisfaction of the desire for profit;

Undertake to husband and cherish Tasmania's living resources so that we do minimum damage to the web of life of which we are part while preventing the extinction or serious depletion of any form of life by our individual, group or communal actions;

And we shall

Create new institutions so that all who wish may participate in making laws and decisions at all levels concerning the social, cultural, political and economic life of the community;

Provide institutions for the peaceful and unimpeded evolution of the community and for the maintenance of justice and equal opportunity for all people;

Change our society and our culture to prevent a tyranny of rationality, at the expense of values, by which we may lose the unique adaptability of our species for meeting cultural and environmental change;

Prevent alienation of people in their social and work roles and functions while making scientific, technical, and vocational knowledge and practice free and open to all;

Create a new community in which men and women shall be valued for their personal skills, for the material and non-material worth of these skills to groups and the whole community, for their service to the community, and for their non-competitive achievement in all aspects of life;

Live as equal members of our society to maintain a community governed by rational, non-sectional law;

Preserve specific areas of private and group life where private thought, speech, and action is of individual or group importance and does not interfere unreasonably with others;

And vest our individual and communal rights in a parliament of representatives chosen by all to enforce our law for as long as that power is not used unfairly to advantage or disadvantage any individual or group in the community.

Appendix 3
Excerpts from the 1989 Tasmanian Parliamentary Accord

Excerpts from the 1989 Tasmanian Parliamentary Accord between the Green Independents and the Parliamentary Labor Party that relate to forestry and national parks

HEADS OF AGREEMENT

6 A Douglas Apsley National Park will be gazetted in 1989 with the boundaries defined by the Department of Lands, Parks and Wildlife by agreement with the Green Independents.

7 The Huon Forest Products venture will not be allowed to proceed.

8 The political mistakes of the Gray Government have ruled out Wesley Vale as the site for a future pulp mill. Therefore, there will be no new pulp mill at Wesley Vale.

9 The State export woodchip quota will not exceed 2.889 million tonnes per annum.

10 The Denison Spires area, Hartz Mountain National Park and Little Fisher Valley will be immediately added to the current World Heritage nomination. The Denison Spires area and the Little Fisher Valley will be gazetted as National Parks in 1989.

11 The following areas will be nominated immediately for World Heritage listing:

- Hartz Mountains National Park
- Little Fisher Valley

while the following areas will be considered for listing as a matter of priority:

- Central Plateau Protected Area and adjacent forest reserves
- the Campbell River area
- the Eldon Range
- lower Gordon River (catchment).

12 The World Heritage Planning Team within the Department of Lands, Parks and Wildlife will prepare a report on the appropriate boundaries of a Western Tasmania World Heritage Area (with the existing National Estate Area as a reference point) for presentation to the World Heritage Committee by 1989.

13 Any National Estate forests within the greater Western Tasmania National Estate Area that a Labor Government agrees to protect will also be nominated for World Heritage.

14 National Estate Forests

This agreement recognises the importance of protecting National Estate forests and the need for a Labor Government to strive to achieve this objective. It will be the stated policy of a Labor Government to give full and on-going protection of National Estate values and, together with the assistance of the Federal Government, ensure that the interests of the timber-industry workers are protected. The Independents will continue to work for the complete protection of Tasmania's National Estate Areas. To achieve the above objectives:

a) Further logging and roading will not proceed in areas in which logging has not already been approved under the Federal–State forestry agreement in order to prevent exploitation of the forest resource pending the outcome of the review.

b) Current or scheduled logging and roading operations in the following areas will not be allowed to proceed:
   i) East Picton;
   ii) Jackeys Marsh;
   iii) Lake Ina.

c) A review process lasting at least a year will be established immediately upon a Labor Government assuming office to:
   i) investigate alternatives to logging in National Estate Areas and nominated areas (as at 31 May 1989) that are not already nominated for World Heritage or included within State Reserves;
   ii) assess the economic and employment effects of protecting those National Estate Areas from logging operations and the strategies available to overcome these effects;
   iii) ensure that the interests of timber-industry workers are protected.

d) That review process will be carried out by a forestry task force whose composition, structure and precise terms of reference will be determined by the Premier, Mr Field, and Dr Bob Brown.

e) The resources, staff, data and facilities of the Forestry Commission will be at the disposal of the task force. The task force will also have access to data on forest resources in company concessions.
f) Representations to the review will be sought from industry, timber-industry unions, logging contractors, independent experts, conservation groups and other relevant parties.
g) The review will investigate but not be limited to:
  i) determination of the proportion of the State's total timber resource that lies within National Estate Areas not already nominated for World Heritage or included within State Reserves;
  ii) installation of sawlog-recovery (flitch) mills at the State's chipmills and pulpmills;
  iii) allocation of the $41.5 million remaining from the 1988 forestry compensation package;
  iv) increased utilisation of the sawlog resource on private land;
  v) better utilisation of the sawlog resource at the State's sawmills;
  vi) intensive management of selected sites;
  vii) improving efficiency by changing the concession system;
  viii) establishment of sawlog plantations;
  ix) increasing financial incentives to logging contractors to recover good-quality sawlogs;
  x) better supervision of logging operations and segregation of sawlogs throughout the State, and particularly in the ANM concession;
  xi) technological developments in wood processing;
  xii) shedding of timber-industry jobs through automation and restructuring carried out by the timber industry itself;
  xiii) the extent to which logging of National Estate Areas can be delayed without loss of jobs in the forestry industry;
  xiv) the role and future of Tasmania's small sawmills in the forestry industry.
h) The $41.5 million remaining from last year's compensation package will be used to help implement alternatives to logging National Estate Areas and to ensure the protection of timber-industry workers' interests. The Independents will participate in the allocation of that $41.5 million.

15 Any National Estate Areas which the review identifies as not essential to the logging industry will be protected as national parks.

16 There will be a full review of the Forestry Commission and a move to abolish the concession system.

17 Areas already agreed to be nominated for World Heritage as a result of last year's Tasmanian Forestry Agreement will be protected as national parks, in 1989.

Notes

Preface

1 Kevin Kiernan, 'Conservation, Timber and Perceived Values at Mt Field, Tasmania' in *Australia's Everchanging Forests V: Proceedings of the Fifth National Conference on Australia's Forest History*, Centre for Resource and Environmental Studies, Australian National University, Canberra, 2002, p. 224.

Introduction

1 Michael Williams, *Deforesting the Earth: From Prehistory to Global Crisis, An Abridgement*, The University of Chicago Press, Chicago, 2006, pp. 370, 382.
2 As quoted on Interpretation panel, Yosemite National Park, California.
3 As quoted in *Gordon River Splits* film, Tasmanian Wilderness Society, Hobart, 1982.

Chapter 1 Growth of the HEC

1 Roger Lupton, *Lifeblood: Tasmania's Hydro Power*, Focus Publishing, Edgecliff, NSW, 2000, p. 204.
2 ibid., pp. 26–31.
3 ibid., pp. 48, 49.
4 ibid., p. 57.
5 ibid., pp. 62, 64.
6 ibid., p. 66.

7 ibid., p. 80.
8 Les Southwell, *The Mountains of Paradise: The Wilderness of South-West Tasmania,* Les Southwell Pty Ltd, Camberwell, Victoria, 1983, p. 16.
9 Doug Lowe, *The Price of Power: The Politics Behind the Tasmanian Dams Case,* Macmillan, Melbourne, 1984, p. 86.
10 Lloyd Robson, *A Short History of Tasmania,* Oxford University Press, Melbourne, 1985, p. 126.
11 Lupton, *Lifeblood,* p. 131.
12 ibid., pp. 151, 152.
13 Southwell, *The Mountains of Paradise,* p. 16.
14 Lupton, *Lifeblood,* p. 175.
15 ibid., p. 10.
16 ibid., p. 176.
17 ibid., p. 177.
18 Keith McKenry, 'A History and Critical Analysis of the Controversy Concerning the Gordon River Power Scheme', in *Pedder Papers: Anatomy of a Decision,* Australian Conservation Foundation, Melbourne, 1972, p. 12.
19 Lupton, *Lifeblood,* p. 198.
20 ibid., p. 209.
21 ibid., p. 210.

Chapter 2 Lake Pedder

1 Bob Brown, *Lake Pedder,* The Wilderness Society, Hobart, 1986, p. 16.
2 John Burton et al., *Lake Pedder Committee of Enquiry: Final Report,* Australian Government Publishing Service, Canberra, 1974, pp. 99, 100.
3 Eric Reece interviewed on ABC Radio National's *Premiers Past* program, 28.12.06.
4 William J. Lines, *Patriots: Defending Australia's Natural Heritage,* University of Queensland Press, St Lucia, Brisbane, 2006, p. 52.
5 Keith McKenry, 'A History and Critical Analysis of the Controversy Concerning the Gordon River Power Scheme', in *Pedder Papers: Anatomy of a Decision,* Australian Conservation Foundation, Melbourne, 1972, p. 15.
6 Burton, *Lake Pedder Committee of Enquiry,* p. 294.
7 Roger Green (ed.), *Battle for the Franklin,* Australian Conservation Foundation/ Fontana Books, Melbourne, 1981, p. 21.
8 Helen Gee (ed.), *The South West Book: A Tasmanian Wilderness,* Australian Conservation Foundation, Melbourne, 1978, p. 277.

9 McKenry, 'A History . . . the Gordon River Power Scheme', p. 16.
10 Cassandra Pybus and Richard Flanagan (eds), *The Rest of the World is Watching: Tasmania and the Greens,* Pan Macmillan, Sydney, 1990, p. 21.
11 Lines, *Patriots,* p. 58.
12 Les Southwell, *The Mountains of Paradise: The Wilderness of South-West Tasmania,* Les Southwell Pty Ltd, Camberwell, Victoria, 1983, p. 33.
13 McKenry, 'A History . . . the Gordon River Power Scheme', p. 16.
14 Pybus and Flanagan, *The Rest of the World is Watching,* p. 22.
15 McKenry, 'A History . . . the Gordon River Power Scheme', p. 17.
16 ibid.
17 ibid., p. 19.
18 Pybus and Flanagan, *The Rest of the World is Watching,* p. 23.
19 McKenry, 'A History . . . the Gordon River Power Scheme', p. 22.
20 Pybus and Flanagan, *The Rest of the World is Watching,* p. 23.
21 Southwell, *The Mountains of Paradise,* p. 17.
22 Roger Lupton, *Lifeblood: Tasmania's Hydro Power,* Focus Publishing, Edgecliff, NSW , 2000, p. 247.
23 Southwell, *Mountains of Paradise,* p. 24.
24 ibid.
25 Burton, *Lake Pedder Committee of Enquiry,* p. 48.
26 *Pedder Papers: Anatomy of a Decision,* Australian Conservation Foundation, Melbourne, 1972, pp. 23, 33.
27 Green, *Battle for the Franklin,* p. 56.
28 Southwell, *Mountains of Paradise,* p. 28.
29 McKenry, 'A History . . . the Gordon River Power Scheme', p. 24.
30 Green, *Battle for the Franklin,* p. 52.
31 Pybus and Flanagan, *The Rest of the World is Watching,* pp. 27, 28.
32 Sam Simpson, 'Flood that Sparked a Green Awakening', *The Sunday Tasmanian,* 7.6.92, p. 14.
33 Green, *Battle for the Franklin,* p. 56.
34 Lupton, *Lifeblood,* p. 261.
35 Burton, *Lake Pedder Committee of Enquiry,* p. 4.
36 Lines, *Patriots,* p. 129.
37 Green, *Battle for the Franklin,* p. 71.
38 Southwell, *Mountains of Paradise,* p. 36.
39 ibid., p. 37.
40 ibid.
41 ibid., p. 38.
42 Green, *Battle for the Franklin,* p. 75.
43 Burton, *Lake Pedder Committee of Enquiry,* p. 310.

44 ibid., p. 309.
45 ibid., p. 316.
46 Brown, *Lake Pedder*, p. 13.
47 Lupton, *Lifeblood*, p. 289.
48 Helen Gee et al., *Reflections*, Issue 2, September 1994, Pedder 2000, Hobart, 1994, downloaded from <http://www.lakepedder.org/resources/reflections/reflect2.htm> in September 2006.
49 Colin Chung, 'Pedder Stays Unplugged', *The Mercury*, 27.6.95, p. 3.
50 *Togatus 5*, Tasmanian University Union, Hobart, 2000, p. 18.
51 Southwell, *Mountains of Paradise*, p. 26.
52 *Lake Pedder* (film), Film Australia, Lindfield, 1997.
53 Simpson, 'Flood that Sparked a Green Awakening'.

Chapter 3 The Franklin and beyond

1 Roger Lupton, *Lifeblood: Tasmania's Hydro Power*, Focus Publishing, Edgecliff, NSW , 2000, pp. 271, 289.
2 Peter Thompson, *Bob Brown of the Franklin River*, Allen & Unwin, Sydney, 1984, p. 51.
3 Roger Green (ed.), *Battle for the Franklin*, Australian Conservation Foundation/Fontana Books, Melbourne, 1981, p. 123.
4 Jerry Fetherston, 'A Voice in the Wilderness', *New Idea*, 26.1.80, p. 10.
5 ibid., p. 125.
6 Green, *Battle for the Franklin*, p. 125.
7 ibid., p. 127.
8 Thompson, *Bob Brown*, p. 114.
9 Les Southwell, *The Mountains of Paradise: The Wilderness of South-West Tasmania*, Les Southwell Pty Ltd, Camberwell, Victoria, 1983, p. 50.
10 Doug Lowe, *The Price of Power: The Politics Behind the Tasmanian Dams Case*, Macmillan, South Melbourne, 1984, p. 116.
11 Bob Connolly, *The Fight for the Franklin: The Story of Australia's Last Wild River*, Cassell, 1981, p. 127.
12 Thompson, *Bob Brown*, p. 118.
13 'Courage, Wisdom, Hailed', *The Mercury*, 12.7.80, p. 3.
14 Doug Lowe interviewed on ABC Radio National's *Premiers Past* program, 2.1.07.
15 William J. Lines, *Patriots: Defending Australia's Natural Heritage*, University of Queensland Press, St Lucia, Brisbane, 2006, p. 197.
16 Doug Lowe, ABC Radio's *Premiers Past*.

17 Southwell, *The Mountains of Paradise,* p. 52.
18 As related by Doug Lowe in a talk given at the University of Tasmania to the University History Society on 29.8.07.
19 ibid.
20 Southwell, *The Mountains of Paradise,* p. 53.
21 Thompson, *Bob Brown,* p. 145.
22 ibid.
23 ibid., p. 154.
24 Lines, *Patriots,* p. 207.
25 Thompson, *Bob Brown,* p. 156.
26 Greg Buckman, *Tasmania—A Small Cork on a Big Economic Ocean: An Overview of the Tasmanian Economy, its Unemployment Problem and Green Policy Options,* Tasmanian Institute for Independent Policy Studies, Hobart, 1989, p. 6.
27 Greg Buckman, 'Focus: Tasmania's Economy', in *Edgeways,* Issue 4, Autumn 1988, Eumarrah Publications, Hobart, 1988, p. 21.
28 Buckman, 'Focus: Tasmania's Economy', p. 20.
29 Bob Burton, *Overpowering Tasmania: A Briefing Paper on Power Demand and Supply,* The Wilderness Society, Hobart, 1985, p. 30.
30 Peter Thompson, *Power in Tasmania,* Australian Conservation Foundation, Melbourne, 1981, pp. 142, 165.
31 Southwell, *The Mountains of Paradise,* p. 54.
32 Thompson, *Bob Brown,* p. 160.
33 The Wilderness Society, *Franklin Blockade by the Blockaders,* The Wilderness Society, Hobart, 1983, p. 10.
34 ibid., p. 11.
35 Southwell, *The Mountains of Paradise,* p. 57.
36 The Wilderness Society, *Franklin Blockade,* p. 11.
37 Green, *Battle for the Franklin,* p. 255.
38 The Wilderness Society, *Franklin Blockade,* p. 9.
39 Lines, *Patriots,* p. 208.
40 The Wilderness Society, *Franklin Blockade,* pp. 257, 259.
41 Green, *Battle for the Franklin,* p. 249.
42 Mike Steketee, 'Just Call Me Responsible', *The Australian,* 8.11.07, p. 15.
43 ABC Radio National, *Hindsight* program, 29.1.06.
44 ibid.
45 The Wilderness Society, *Franklin Blockade,* p. 11.
46 ibid., p. 12.
47 ibid.
48 ibid.
49 ibid., p. 13.

50 Southwell, *The Mountains of Paradise*, p. 62.
51 Gregg Borschmann, 'Win Gives Brown Hope for the "Big Problems"', *The Age*, 2.7.83, p. 5.
52 ibid.
53 Southwell, *The Mountains of Paradise*, p. 63.
54 Lupton, *Lifeblood*, p. 337.
55 ibid., p. 383.
56 ibid., pp. 385, 407.
57 Senator Bob Brown, 'National Press Club Address 25 September 2001', downloaded from <www.bobbrown.org.au> in September 2006.

Chapter 4 Sawmilling to industrial forestry

1 Helen Gee (ed.), *For the Forests: A History of the Tasmanian Forests Campaigns*, The Wilderness Society, Hobart, 2001, p. vii.
2 ibid., p. 351.
3 ibid., p. viii.
4 Phillip Hoysted, *The Content and Historical Development of Forestry Legislation in Tasmania*, Centre for Environmental Studies, University of Tasmania, Hobart, 1983, p. 3.
5 Gee, *For the Forests*, pp. viii, ix.
6 ibid., p. ix.
7 L.T. Curran, *A History of Forestry in Australia*, Australian National University Press, Canberra, 1985, p. 67.
8 ibid., p. 75.
9 Hoysted, *The Content . . . of Forestry Legislation*, p. 33.
10 ibid., p. 4.
11 Gee, *For the Forests*, p. 3.
12 ibid.
13 ibid.
14 Kevin Kiernan, 'Conservation, Timber and Perceived Values at Mt Field, Tasmania' in *Australia's Everchanging Forests V: Proceedings of the Fifth National Conference on Australia's Forest History*, Centre for Resource and Environmental Studies, Australian National University, Canberra, 2002, p. 223.
15 Gee, *For the Forests*, p. 4.
16 ibid., p. 339.
17 ibid., p. xi.

18 David Mercer and Jim Petersen, 'The Revocation of National Parks and Equivalent Reserves in Tasmania', *Search,* Volume 17, Numbers 5–6, May–June 1986, p. 138.
19 Curran, *A History of Forestry,* p. 72.
20 ibid., p. 74.
21 ibid., p. 80.
22 ibid., p. 81.
23 ibid., p. 83.
24 Marcus Priest, 'Tasmania Once Again Logs Into Federal Scene', *Australian Financial Review,* 24.7.07, p. 60.
25 R. and V. Routley, *The Fight for the Forests: The Takeover of Australian Forests for Pines, Wood Chips and Intensive Forestry,* Research School of Social Sciences, Australian National University, Canberra, 1975, p. 108.
26 ibid.
27 Curran, *A History of Forestry,* p. 87.
28 Helen Gee and Janet Fenton (eds), *The South West Book: A Tasmanian Wilderness,* Australian Conservation Foundation, Melbourne, 1979, p. 195.
29 Gee, *For the Forests,* p. 84.
30 ibid., p. 187.
31 ibid., p. 202.
32 ibid.
33 Max Angus (ed.), *The World of Olegas Truchanas,* OBM Pty Ltd, Hobart, 1975, p. 41.
34 Gee, *For the Forests,* p. 17.
35 ibid., p. 17.
36 ibid., p. 215.

Chapter 5 Farmhouse Creek to the Salamanca Agreement

1 Helen Gee (ed.), *For the Forests: A History of the Tasmanian Forests Campaigns,* The Wilderness Society, Hobart, 2001, p. 203.
2 Drew Hutton and Libby Connors, *A History of the Australian Environment Movement,* Cambridge University Press, Melbourne, 1999, p. 180.
3 William J. Lines, *Patriots: Defending Australia's Natural Heritage,* University of Queensland Press, St Lucia, Brisbane, 2006, p. 238.
4 Hutton and Connors, *Australian Environment Movement,* p. 210.

5 Debbie Quarmby, The Politics of Parks: A History of Tasmania's National Parks 1885–2005, PhD thesis, Murdoch University, 2006, p. 250.
6 Hutton and Connors, *Australian Environment Movement,* p. 353.
7 ibid., p. 183.
8 Gee, *For the Forests,* p. 230.
9 ibid., p. 170.
10 ibid., p. 171.
11 ibid.
12 ibid., p. 172.
13 *Multinational Monitor,* October 1987, p. 18, as reported in Lines, *Patriots,* p. 240.
14 Gee, *For the Forests,* p. 216.
15 ibid., p. 354.
16 Hutton and Connors, *Australian Environment Movement,* p. 182.
17 Louise Mendel, Scenery to Wilderness: National Park Development in Tasmania, 1916–1992, PhD thesis, University of Tasmania, 1999, p. 175.
18 Gee, *For the Forests,* p. 221.
19 ibid., p. 355.
20 ibid., p. 227.
21 Hutton and Connors, *Australian Environment Movement,* p. 187.
22 Quarmby, The Politics of Parks, p. 252.
23 Mendel, Scenery to Wilderness, p. 177.
24 Gee, *For the Forests,* p. 355.
25 Hutton and Connors, *Australian Environment Movement,* p. 204.
26 Cassandra Pybus and Richard Flanagan (eds), *The Rest of the World is Watching: Tasmania and the Greens,* Pan Macmillan, Sydney, 1990, p. 54.
27 ibid., p. 57.
28 Hutton and Connors, *Australian Environment Movement,* p. 206.
29 Gee, *For the Forests,* p. 355.
30 Lines, *Patriots,* p. 257.
31 Gee, *For the Forests,* p. 185.
32 *The Green–ALP Accord,* The Green Independents, Hobart, 1989, pp. 9, 10.
33 Michael Field interviewed on *Premiers Past,* ABC Radio National, 12.1.07.
34 Gee, *For the Forests,* p. 142.
35 ibid., p. 144.
36 ibid., p. 356.
37 ibid., p. 357.
38 Michael Field interviewed on *Premiers Past.*
39 Gee, *For the Forests,* p. 300.

Chapter 6 The Regional and Lennon–Howard forest agreements

1 Helen Gee (ed.), *For the Forests: A History of the Tasmanian Forests Campaigns*, The Wilderness Society, Hobart, 2001, p. 303.
2 ibid., p. 315.
3 Judith Ajani, *The Forest Wars*, Melbourne University Press, Carlton, Melbourne, 2007, p. 7.
4 Gee, *For the Forests*, p. 316.
5 ibid., pp. 362, 363.
6 Geoff Law, 'What's Wrong With the Regional Forest Agreement?', The Wilderness Society, Hobart, 2002, p. 1, downloaded from <www.wilderness.org> in 2006.
7 Geoff Law, 'A Proposal for a Styx Valley of the Giants National Park', The Wilderness Society, 2001, p. 2, downloaded from <www.wilderness.org> in 2006.
8 Law, 'A Proposal for a Styx Valley of the Giants National Park', p. 22.
9 Debbie Quarmby, The Politics of Parks: A History of Tasmania's National Parks 1885–2005, PhD thesis, Murdoch University, 2006, p. 259.
10 Amanda Sully, 'Conservationists Outraged at Barbaric and Dishonest Moves by Labor to Log Deferred Forests', The Wilderness Society, Hobart, 1998, p. 1, downloaded from <www.wilderness.org> in 2006.
11 Gee, *For the Forests*, p. 360.
12 Geoff Law, 'Tasmania to Log an Area of Public Land Larger than Hobart's Urban Area This Year', The Wilderness Society, Hobart, 2000, p. 1, downloaded from <www.wilderness.org> in 2006.
13 Geoff Law, 'Statistics on the Tasmanian Forests Debate', The Wilderness Society, Hobart, 2002, p. 1, downloaded from <www.wilderness.org> in 2006.
14 Geoff Law, 'Forestry Tasmania's Atrocious Environmental, Economic and Social Record', The Wilderness Society, Hobart, 2002, p. 1, downloaded from <www.wilderness.org> in 2006.
15 Law, 'Statistics on the Tasmanian Forests Debate', p. 1.
16 ibid., p. 4.
17 Danny Rose, 'Poll-axed: Forest Survey has Lennon Spitting Chips', *The Mercury*, 29.1.04, p. 1.
18 Law, 'Statistics on the Tasmanian Forests Debate', p. 2.
19 ibid.

20 Quoted in film, *Tasmania's Clean Green Future: Too Precious to Pulp?*, produced by The Wilderness Society, Hobart, in 2007.
21 Law, 'What's Wrong With the Regional Forest Agreement?', p. 1.
22 Suzi Pipes, *Tasmanian Forests Fact File June 2004*, The Wilderness Society, Hobart, 2004, p. 1.
23 Peter Thompson, *Power in Tasmania*, Australian Conservation Foundation, Melbourne, 1981, p. 152.
24 Randal O'Toole, *Economic Review of the Forestry Commission of Tasmania*, CHEC, Eugene, Oregon, 1989, p. 1.
25 Based on annual reports published by Gunns and Forestry Tasmania for 2000–01 to 2004–05.
26 Timber Workers for Forests, 'Tasmania's Timber Industry Jobs: Briefing Sheet September 2004', downloaded from <www.twff.com> in 2007.
27 Ellen Whinnett, 'Latham Takes the Middle Path', *The Mercury*, 19.3.04, p. 4.
28 Sid Maher and Dennis Shanahan, 'PM Moves to Protect Old Forests', *The Weekend Australian*, 4.9.04, p. 6.
29 Chris Johnson, 'Loggers Cheer Howard', *The Examiner*, 7.9.04, p. 1.
30 Geoff Law, *Analysis of the May 2005 Howard–Lennon Forests Agreement*, The Wilderness Society, Hobart, 2005, p. 11.
31 ibid.
32 The Wilderness Society, 'Gov't Tassie Forest Plan Won't End Old-growth Logging', Hobart, 2005, p. 2, downloaded from <www.wilderness.org> in 2006.
33 Michelle Paine, 'Greens Call it Hit and Myth', *The Saturday Mercury*, 14.5.05, p. 3.
34 Philippa Duncan, 'Green Joy at Legal Victory', *The Mercury*, 20.12.06, p. 1.

Chapter 7 Gunns

1 Judith Ajani, *The Forest Wars*, Melbourne University Press, Carlton, Melbourne, 2007, p. 285.
2 ibid., pp. 285, 286.
3 ibid.
4 Simon Bevilacqua, 'Tassie's Great Bribe Scandal', *The Sunday Tasmanian*, 28.6.02, pp. 6, 7.
5 ibid.
6 'Gunns is Big Government Money Donor', *The Examiner*, 3.2.04, p. 3.

7 The Wilderness Society, 'Fightback against woodchip giants $6.4 million lawsuit against conservationists begins', 14.1.05, downloaded from <www.wilderness.org> in 2006.

8 Gavin Lower, 'New Blow for Gunns', *The Mercury*, 21.10.06, pp. 1, 2.

9 Gavin Lower, 'Gunns to Continue Legal Fight', *The Mercury*, 29.8.06, p. 9.

10 ibid.

11 ibid.

12 Gavin Lower, 'Gunns Focuses Claim on 14', *The Mercury*, 4.4.07, p. 5.

13 Advertisement by Gunns: 'Bell Bay Pulp Mill Project Update: The Facts—Wood Supply', *The Mercury*, 14.4.07, p. 11.

14 Sue Neales, 'Carry on Chipping: Exports Set to Almost Double Despite Planned Pulp Mill', *The Mercury*, 4.8.06, p. 1.

15 ibid.

16 Mathew Denholm, 'Pulp Mill to Feed on Native Forests', *The Australian*, 11.7.06, p. 5.

17 Jennifer Hewitt, 'How Turnbull's Plan was Beaten to a Pulp', *The Australian*, 1.9.07, p. 4.

18 ABC Radio, 25.6.06, 'AMA Criticises Tasmanian Mill Proposal', as quoted in TWS pamphlet: *What Will the Pulp Mill do to Our Quality of Life?*, The Wilderness Society, Hobart, 2007.

19 Anne Mather, 'Gay Slams Mill Fears', *The Mercury*, 6.11.06, p. 3.

20 Simon Bevilacqua, 'Mill Blow: Threat to Pollution Treaty, Says Expert', *Sunday Tasmanian*, 25.3.07, p. 1.

21 Philippa Duncan, 'Gunns' Odour Case on the Nose', *The Mercury*, 14.11.06, p. 10.

22 The Wilderness Society, *Pulp Mill Campaign Update*, Hobart, 2007, p. 3.

23 Investors for the Future of Tasmania advertisement: 'Every Kraft Pulp Mill Smells', *The Mercury*, 28.7.07, p. 26.

24 Margaret Blakers, 'Green Institute Backgrounder: Gunns Proposed Pulpmill, Greenhouse Gas Emissions', downloaded from <www.christine milne.org.au> in 2007.

25 The Wilderness Society, *What Will the Pulp Mill do to Our Quality of Life?* pamphlet, Hobart, 2007.

26 ibid.

27 Sue Neales, 'Fisher Fury As Mill Dioxin Fear Surfaces', *The Mercury*, 31.5.07, p. 9.

28 Mathew Denholm, 'Gunns Go-ahead Despite Failures', *The Australian*, 6.7.07, p. 5.

29 Sue Neales, 'Death, Jobs Lost in Mill's Hidden Costs', *The Mercury*, 23.8.07, pp. 4, 5.
30 Charles Waterhouse, 'Putt Fires Up Pulp-Mill Row', *The Sunday Tasmanian*, 7.1.07, p. 2.
31 Sue Neales, 'Gay Threat to Axe Mill', *The Mercury*, 10.1.07, p. 1.
32 Sue Neales, 'Pulp Mill Approval Alarm', *The Mercury*, 13.1.07, p. 2.
33 Philippa Duncan, 'Mill Backflip Fury', *The Mercury*, 1.2.07, p. 3.
34 Nick Clark, 'Timeline Accepted by Gunns', *The Mercury*, 23.2.07, p. 3.
35 Sue Neales, 'Lennon Push to Speed Up Mill', *The Mercury*, 14.3.07, p. 7.
36 Philippa Duncan, 'Pulp Claims Rock Premier', *The Mercury*, 21.3.07, p. 1.
37 Interview on *4 Corners*, ABC TV, 30.7.07 as reported in Michelle Paine, 'Gunns Gives Former Judge Both Barrels', *The Mercury*, 31.7.07, p. 5.
38 Nick Clark, 'Gunns Turns Up Heat', *The Mercury*, 15.3.07, p. 2.
39 Sue Neales, 'Shock Mill Revelations', *The Mercury*, 7.6.07, p. 11 and Mathew Denholm, 'Gunns Forewarned on Mill Bid Failure', *The Australian*, 15.6.07, p. 6.
40 'New Law to Fast-track Mill', *The Australian*, 20.3.07, p. 6.
41 Philippa Duncan, 'Mill "Leverage" Riddle', *The Mercury*, 3.4.07, p. 1.
42 Nick Clark, 'Poll Pulps Claim Mill is Popular', *The Mercury*, 8.5.07, p. 7.
43 Simon Bevilacqua, 'Assessor of Mill Has Links to Builders', *The Sunday Tasmanian*, 22.4.07, p. 7.
44 Philippa Duncan, 'Flaws in Gunns' Tamar Pulp Plan', *The Mercury*, 6.7.07, p. 4.
45 Investors for the Future of Tasmania advertisement: 'Eric, Guy and Team . . .', *The Mercury*, 21.7.07, p. 21.
46 The Wilderness Society, *Pulp Mill Campaign Update*, p. 2.
47 Mathew Denholm, 'Green Groups Object to Pulp Mill Spinner', *The Australian*, 4.7.07, p. 5.
48 Sue Neales, 'Pulp Mill Survey Shock: Fast-Track Opposed by Two-thirds', *The Mercury*, 9.8.07, p. 9.
49 Luke Sayer, 'Lennon Looks to Healing Divisions in Community', *The Mercury*, 5.10.07, p. 3.
50 Luke Sayer, 'Confidence Boost Tempered by Health Doubts', *The Mercury*, 5.10.07, p. 6.
51 ABC on-line website story, 'Brown Accuses Garrett of Deserting Environment', downloaded from <www.abc.net.au> on 5.10.07.
52 Tim Martain, 'Garrett Faces Protest', *The Sunday Tasmanian*, 2.9.07, p. 4.
53 Mathew Warren, 'Energy Offset to Keep Mill Neutral', *The Australian*, 5.10.07, p. 4.

54 James Button, 'Swedes Cast Doubt on Mill Standards', *The Age,* 15.10.07, p. 2.
55 Geoff Law, *New Pulp-Mill Poll Highlights Parties' Failure on Climate Change,* The Wilderness Society, Hobart, 12.11.07, p. 1.
56 Cassandra Pybus and Richard Flanagan (eds), *The Rest of the World is Watching: Tasmania and the Greens,* Pan Macmillan, Sydney, 1990, p. 6.
57 David Killick, 'The Wilderness Society has Promised "Generations of Conflict" if the Tamar Valley Pulp Mill Goes Ahead', *The Mercury,* 12.2.08.
58 Mathew Denholm, 'Thousands Sign for Pulpmill Protests', *The Australian* (on-line version), 13.2.08.
59 James Kirkby, 'Gunns Might Yet Find the Miller's Tale Has an Unhappy Ending', *The Sunday Age,* 18.11.07, p. 20.
60 Ajani, *The Forest Wars,* pp. 142, 273, 276.

Chapter 8 Mount Lyell to Benders Quarry

1 Alison Alexander (ed.), *The Companion to Tasmanian History,* Centre for Tasmanian Historical Studies, University of Tasmania, Hobart, 2005, p. 337.
2 Patsy Crawford, *King: The Story of a River,* Montpelier Press, Hobart, 2000, p. 46.
3 ibid., pp. 56, 63.
4 ibid., p. 69.
5 ibid., p. 65.
6 Geoffrey Blainey, *The Peaks of Lyell,* St David's Park Publishing, Hobart, 1993, p. 99.
7 Crawford, *King,* p. 118.
8 ibid., p. 120.
9 ibid., p. 70.
10 ibid., pp. 122, 123.
11 Bob Burton, *Undermining Tasmania: A Research Paper on the Mining Industry,* The Wilderness Society, Hobart, 1989, pp. 3–5.
12 Peter Thompson, *Power in Tasmania,* Australian Conservation Foundation, Melbourne, 1981, p. 154.
13 Christobel Mattingley, *King of the Wilderness: The Life of Deny King,* Text Publishing, Melbourne, 2001, p. 40.
14 ibid., pp. 160, 258.

15 ibid., pp. 182, 307.
16 ibid., p. 298.
17 Personal correspondence with Helen Gee, Friends of Melaleuca organisation, 31.7.07.
18 Helen Gee and Janet Fenton (eds), *The South West Book: A Tasmanian Wilderness*, Australian Conservation Foundation, Melbourne, 1979, p. 52.
19 ibid.
20 Gerard Castles, Handcuffed Volunteers: A History of the Scenery Preservation Board in Tasmania 1915–1971, BA honours thesis, University of Tasmania, 1986, p. 48.
21 ibid.
22 Bob Burton, *Undermining Tasmania*, p. 42.
23 ibid.
24 ibid., p. 43.
25 Keith Vallance, 'President's Annual Report', *Tasmanian Conservation Trust Circular, September 1973*, Tasmanian Conservation Trust, Hobart, 1973, p. v.
26 Pat Wessing, 'The Precipitous Bluff Case' in Helen Gee (ed.), *The South West Book*, p. 265.
27 Keith Vallance, 'President's Report', *Tasmanian Conservation Trust Circular, April 1975*, Tasmanian Conservation Trust, Hobart, 1975, p. 2.
28 Bruce Davis, 'President's Report', *Tasmanian Conservation Trust Circular, July/August 1977*, Tasmanian Conservation Trust, Hobart, 1977, p. 2.
29 Official correspondence from Michael Field to Bob Hawke, 11.9.89, p. 2.

Chapter 9 The Scenery Preservation Board

1 Gerard Castles, Handcuffed Volunteers: A History of the Scenery Preservation Board in Tasmania 1915–1971, BA honours thesis, University of Tasmania, 1986, p. 83.
2 Debbie Quarmby, The Politics of Parks: A History of Tasmania's National Parks 1885–2005, PhD thesis, Murdoch University, 2006, p. 35.
3 ibid., p. 36.
4 ibid., pp. 46, 63.
5 ibid., p. 59.
6 ibid., p. 72.
7 Louise Mendel, Scenery to Wilderness: National Park Development in Tasmania, 1916–1992, PhD thesis, University of Tasmania Doctor of Philosophy degree, 1999, p. 24.

8 Quarmby, The Politics of Parks, pp. 74, 75.
9 ibid., pp. 2, 46.
10 ibid., p. 54.
11 Castles, Handcuffed Volunteers, p. 23.
12 ibid., p. 32.
13 ibid., p. 36.
14 ibid.
15 Quarmby, The Politics of Parks, p. 85.
16 Ron Sutton, Tourism in National Parks: Managing a Paradoxical Mandate, Masters thesis, University of Tasmania, Master of Environmental Management degree, 1994, p. 7.
17 Roger and Carol Shively Anderson, *Yellowstone: The Story Behind the Scenery*, KC Publications, Las Vegas, 2006, p. 41.
18 ibid., p. 44.
19 ibid.
20 Quarmby, The Politics of Parks, p. 55.
21 ibid., p. 79.
22 ibid., p. 81.
23 Mendel, Scenery to Wilderness, p. 109
24 Quarmby, The Politics of Parks, pp. 86, 87.
25 ibid., p. 88.
26 ibid., pp. 89, 90.
27 ibid., p. 94.
28 ibid., pp. 76–8.
29 ibid., p. 111.
30 Margaret Giordano, *A Man and a Mountain: The Story of Gustav Weindorfer*, Regal Publications, Launceston, 1987, pp. 2, 7, 14.
31 ibid., p. 31.
32 Quarmby, The Politics of Parks, p. 113.
33 Castles, Handcuffed Volunteers, p. 48.
34 Quarmby, The Politics of Parks, p. 133.
35 Mendel, Scenery to Wilderness, p. 100.
36 Castles, Handcuffed Volunteers, p. 66.
37 Helen Gee (ed.), *For the Forests: A History of the Tasmanian Forests Campaigns*, The Wilderness Society, Hobart, 2001, p. 339.
38 Mendel, Scenery to Wilderness, p. 63.
39 ibid., p. 64.
40 ibid.
41 ibid., p. 73.
42 ibid., p. 75.
43 ibid., p. 104.

44 Quarmby, The Politics of Parks, p. 171.

45 ibid.

46 'Visitors Guide to Tasmania's National Parks—Strzelecki National Park', Parks and Wildlife Service, downloaded from <www.parks.tas.gov.au> in March 2007.

47 Australian Government Department of the Environment and Heritage, *Australia's World Heritage: Australia's Places of Outstanding Universal Value*, Australian Government, Canberra, 2006, p. 58.

48 Quarmby, The Politics of Parks, p. 174.

49 ibid.

50 ibid., pp. 174, 175.

51 Keith McKenry, 'A History and Critical Analysis of the Controversy Concerning the Gordon River Power Scheme', in *Pedder Papers: Anatomy of a Decision*, Australian Conservation Foundation, Melbourne, 1972, p. 12.

52 Castles, Handcuffed Volunteers, p. 84.

53 Quarmby, The Politics of Parks, p. 169.

54 Mendel, Scenery to Wilderness, p. 80.

55 Quarmby, The Politics of Parks, p. 175.

56 Mendel, Scenery to Wilderness, p. 83.

57 Castles, Handcuffed Volunteers, p. 90.

58 ibid., p. 92.

59 Quarmby, The Politics of Parks, p. 181.

60 Castles, Handcuffed Volunteers, p. 94.

61 Quarmby, The Politics of Parks, p. 182.

62 Mendel, Scenery to Wilderness, p. 107.

63 David Mercer and Jim Peterson, 'The Revocation of National Parks and Equivalent Reserves in Tasmania', *Search*, Volume 17, Numbers 5–6, May–June 1986, p. 138.

Chapter 10 The National Parks and Wildlife Service

1 Debbie Quarmby, The Politics of Parks: A History of Tasmania's National Parks 1885–2005, PhD thesis, Murdoch University, 2006, p. 163.

2 Louise Mendel, Scenery to Wilderness: National Park Development in Tasmania, 1916–1992, PhD thesis, University of Tasmania, 1999, p. 118.

3 Quarmby, The Politics of Parks, p. 165.

4 ibid., p. 165.
5 Mendel, Scenery to Wilderness, p. 133.
6 Quarmby, The Politics of Parks, p. 201.
7 Mendel, Scenery to Wilderness, p. 180.
8 ibid., p. 181.
9 Quarmby, The Politics of Parks, p. 170.
10 ibid., p. 184.
11 Quarmby, The Politics of Parks, p. 225.
12 ibid.
13 Les Southwell, *The Mountains of Paradise: The Wilderness of South-West Tasmania,* Les Southwell Pty Ltd, Camberwell, Victoria, 1983, p. 48.
14 B.W. Davis and H.M. Gee, 'The South West Advisory Committee' in Helen Gee and Janet Fenton (eds), *The South West Book: A Tasmanian Wilderness,* Australian Conservation Foundation, Melbourne, 1979, p. 256.
15 Mendel, Scenery to Wilderness, pp. 157, 159.
16 ibid., pp. 154, 156.
17 Southwell, *The Mountains of Paradise,* p. 48.
18 Quarmby, The Politics of Parks, p. 229.
19 ibid., p. 231.
20 Mendel, Scenery to Wilderness, p. 161.
21 Quarmby, The Politics of Parks, p. 233.
22 Doug Lowe, *The Price of Power: The Politics Behind the Tasmanian Dams Case,* Macmillan Company, South Melbourne, 1984, p. 144.
23 Quarmby, The Politics of Parks, p. 240.
24 ibid., p. 241.
25 ibid., p. 279.
26 ibid.
27 ibid., p. 299.
28 ibid., p. 300.
29 ibid., p. 301.
30 ibid., p. 302.
31 Mendel, Scenery to Wilderness, pp. 185–7.
32 Quarmby, The Politics of Parks, p. 255.
33 ibid., p. 257.
34 ibid., p. 259.
35 ibid.
36 ibid., pp. 261, 262.
37 ibid., p. 262.
38 ibid., p. 297.
39 ibid.

40 ibid.
41 ibid., p. 298.
42 ibid., p. 301.
43 ibid., p. 265.
44 ibid.
45 ibid.

Chapter 11 Conclusion

1 Matthew Newton, 'The River' in *Adventure Journal 2007: Australia's Annual Magazine of Adventure and Exploration*, p. 40.
2 Richard Flanagan, 'Out of Control: The Tragedy of Tasmania's Forests', *The Monthly*, May, 2007, p. 23.
3 ibid., p. 31.
4 Marcus Priest, 'Pulp Friction: Mill Battle Exposes Deep Political Fault Line', *The Australian Financial Review*, 30.7.07, pp. 60, 61.
5 As quoted in Sue Neales, 'Not a Good Look', *The Mercury*, 25.8.07, p. 23.

Index